Human Fatigue Risk Management

Human Fatigue Risk Management
Improving Safety in the Chemical Processing Industry

Susan L. Murray, PhD, PE
Missouri University of Science and Technology
Rolla, MO, United States

Matthew S. Thimgan, PhD
Missouri University of Science and Technology
Rolla, MO, United States

ELSEVIER

AMSTERDAM • BOSTON • HEIDELBERG • LONDON
NEW YORK • OXFORD • PARIS • SAN DIEGO
SAN FRANCISCO • SINGAPORE • SYDNEY • TOKYO

Academic Press is an imprint of Elsevier

Library of Congress Cataloging-in-Publication Data
A catalog record for this book is available from the Library of Congress

British Library Cataloguing-in-Publication Data
A catalogue record for this book is available from the British Library

ISBN: 978-0-12-802412-6

For information on all Academic Press publications
visit our website at https://www.elsevier.com/

Working together
to grow libraries in
developing countries

www.elsevier.com • www.bookaid.org

Publisher: Joe Hayton
Acquisition Editor: Fiona Geraghty
Editorial Project Manager: Lindsay Lawrence
Production Project Manager: Caroline Johnson
Designer: Matthew Limbert

Typeset by Thomson Digital

Contents

About the Authors

Dr Susan Murray is the Interim Chair of the Psychological Science Department and a Professor of Engineering Management & Systems Engineering at the Missouri University of Science and Technology. She earned her doctorate in Industrial Engineering from Texas A&M University. Her research and teaching interests include human factors, safety, process improvement, usability, and improving higher education. Prior to her academic position, she worked in the aerospace industry including 2 years at NASA's Kennedy Space Center. Her goal continues to be making improvements to benefit workers and students.

Dr Matthew Thimgan runs the Sleep Biology Laboratory at the Missouri University of Science and Technology. He earned his doctorate in Cell and Molecular Physiology from the University of North Carolina, Chapel Hill. He continued his training in sleep biology at Washington University in St. Louis. He currently studies sleep in both the human and the fruit fly. In humans, he and members of his lab are identifying biomarkers of sleepiness. In the fruit fly, they are pursuing the genes and biochemical pathways that regulate sleep and wakefulness. His scientific contributions have appeared in general biology journals, book chapters, as well as journals for the sleep field.

Foreword

At the outset, I must say that this publication by Dr Susan Murray and Matthew Thimgan on, "*Human Fatigue Risk Management*," is a very timely and much needed treatise. It is even more important given the need for process safety and sustainable development; a rational and constructive approach to human factors; and the significance of human factors in risk analysis, safety performance, and occurrence of catastrophic incidents.

Worker fatigue is a risk factor and should be managed to prevent catastrophic incidents and improve safety performance. However, this area is not well studied and management systems often do not take into account the increasing complexity of chemical processes, interdependent chemical infrastructure, and the need for considering diverse and competing issues. This publication by Murray and Thimgan delve deeply into all the issues that come into play for human fatigue and innovative approaches to managing the risk.

Murray and Thimgan have been successful in taking a refreshing and poignant look at human fatigue and risk management; they have applied a thoughtful approach in a holistic manner for the analysis and control of risks inherent to human fatigue. The book starts with a very succinct description of the consequences of lack of human fatigue risk management, and then goes into extensive detail about all the factors that are important in assessing fatigue, the impact of fatigue on human performance and incidents, and finally ends up with a very constructive management system for human fatigue. In addition, descriptions of regulatory requirements for fatigue risk management are provided in an easy-to-understand language. The book also provides a very comprehensive review of methods used over the years to understand and manage human fatigue. I sincerely believe that the book has opened up a new vista and perspective on methodological improvements, necessary in the ever-increasing complexity of safe manufacturing and distribution in a competitive world. This book is a must read for process safety and risk-management professionals, human factors specialists, managers, and leaders who want to understand the underlying issues related to human fatigue risk management and effective approaches to dealing with these issues.

M. Sam Mannan
Regents Professor and Executive Director,
Mary Kay O'Connor Process Safety Center,
Texas A&M University,
College Station, TX, United States

Acknowledgments

The authors wish to acknowledge and thank the individuals at Elsevier whose technical expertise made this book possible. In particular, we are in debt to Lindsay Lawrence, Fiona Geraghty, and Caroline Johnson without whom this book would still be just an idea.

Without Dr. Sam Mannon's encouragement and support to include human factors in the effort to improve process safety we would not have even started working in the area of human fatigue risk management. We also thank the reviewers that provided useful feedback on the book proposal.

Thanks to all of the researchers that have dedicated their talents to understanding the impact of adequate sleep on our health and performance as well as how to best incorporate these lessons to improve our lives. Also, a debt of gratitude is owed to the subjects that have allowed themselves to be studied to obtain this important information.

On the personal level we would like to thank our families for their support and understanding while we worked on this book. Without you: Julie, Jack & Dace and Katie, Andrew & Marcus this book would not have been possible.

From Matt:

I need to thank Susan Murray for her guidance and effort in pushing this book forward. She has used all of her experience and expertise to develop the final manuscript. She has been patient and encouraging with this naïve author.

From Susan:

On an individual level, I need to thank Matt Thimgan. Without you I would still be struggling with the biology of sleep. I look up to you not just because you are 15 in. taller than I am but for your expertise and passion as a sleep researcher.

Finally, I need to acknowledge my mother. She died during the writing of this book, which made me realize how much support, encouragement, and love she had given me my entire life. I miss you dearly.

Chapter 1

The consequences of fatigue in the process industries

1.1 BP TEXAS CITY

On Mar. 23, 2005 the BP Texas City Refinery suffered an industrial accident that killed 15 people, injured another 180, and resulted in financial losses exceeding $1.5 billion [1]. The incident occurred when the raffinate splitter tower in the isomerization (ISOM) unit was overfilled with a flammable liquid hydrocarbon. The chain of events leading to this deadly accident spanned several hours. Operational procedures were violated. Critical alarms and control instrumentation provided false indications that failed to alert the operators of the overfill situation. The refinery control room was understaffed and those who were there were exhausted and thoroughly sleep-deprived. When a pressure relief device released flammables there was a lethal series of explosions and fires.

Due to the significance of the disaster, the US Chemical Safety and Hazard Investigation Board (CSB) investigated the BP Texas City facility (the third-largest oil refinery in the United States), management at BP's corporate level, the effectiveness of the Occupational Safety and Health Administration (OSHA), and the industry as a whole. The CSB concluded that the Texas City disaster was due to organizational, safety culture, and human factors at varying corporate levels. Repeated warning signs had been present for years, yet steps were not taken by BP to effectively prevent the tragedy that ultimately happened. The serious safety culture deficiencies were further documented when the refinery experienced two additional serious incidents just a few months after the 2005 explosion. A pipe failure caused a reported $30 million in damage in one accident and the other resulted in a $2 million property loss. In each incident, community shelter-in-place orders were issued [1]. The CSB's 2007 Final Report included a strong focus on causes beyond faulty equipment and operator errors. It was an admonition for the processing industry to improve safety by understanding the human element and to consider workers' limitations.

■ **FIGURE 1.1 Photo of BP Texas City after the accident.** (a) From the CSB website. (b) From the final report. *(Sources: Part a: http://www.csb.gov/bp-texas-city-investigative-photos/. Part b: http://www.csb.gov/bp-america-refinery-explosion/.)*

The CSB's Final Report contained the one of the strongest analyses of human factors as an industrial accident cause. It explored the connections between human fatigue, human performance, and industrial safety in a very detailed fashion. The report established that the individuals working at the time of the accident were clearly severely sleep-deprived and that fatigue risk management was a safety issue that needed to be addressed in the chemical processing industry (Fig. 1.1).

1.2 HUMAN FACTORS AND THE BP TEXAS CITY ACCIDENT

During normal operations a total of four crews worked rotating 12-h shifts at the BP Texas City Plant. Prior to the accident, the ISOM unit was shut down and operators were split into two crews working 12-h shifts during the turnaround operations [1]. On the day of the accident, the day board operator was likely experiencing both acute sleep loss and cumulative sleep debt. He had worked 12-h shifts for 29 consecutive days and generally slept 5–6 h per 24-h period. The day lead operator—who was overloaded training two new operators, dealing with contractors, and working on the ISOM turnaround—had been on duty for 37 consecutive days without a day off prior to the accident. The crew members had a significant commute to and from the refinery. It was common for them to only get 5–6 h of sleep a night.

Fatigue can increase errors, delay reactions, and hamper decision-making [2]. The CSB concluded that fatigue caused by sleep deprivation among the operators working that day degraded the cognitive abilities and performance in solving the overfilling situation. The report [1] stated:

> *In the hours preceding the incident, the tower experienced multiple pressure spikes. In each instance, operators focused on reducing*

pressure: they tried to relieve pressure, but did not effectively question why the pressure spikes were occurring. They were fixated on the symptom of the problem, not the underlying cause and, therefore, did not diagnose the real problem (tower overfill).

Tower overfill was not discussed by the operators during their troubleshooting before the explosion. This type of focused attention to the exclusion of other critical information is called cognitive tunneling and is a common effect of fatigue.

Another key finding from the report was that BP has no fatigue prevention policy or regulations for operators. The contract between the United Steelworkers Union and BP provides a minimum number of hours per work week requirement, but not a maximum. According to BP, "operators were expected to work" the 12-h, 7-days-a-week turnaround schedule, although they were allowed time off if they had scheduled vacation, used personal/vacation time, or had extenuating circumstances that would be considered on a "case-by-case" basis. The company did have fatigue prevention policies that addressed motor vehicle transportation. The BP document states that when multiple fatigue factors are present, a strong argument can be made that fatigue contributes to accidents [3].

FATIGUE RISK MANAGEMENT IN TRANSPORTATION
Hours of service rules have a long history in the transportation industry. Rules for truck drivers were originally established by the federal government around 1939. These rules remained in place, virtually unchanged for decades. The rules were based on consensus rather than science.

OSHA does not have regulations on fatigue prevention that applies to the chemical process industry or oil refinery workers. Other industries in the United States such as nuclear and transportation have government regulations concerning shift work. Many safety professionals joined the CSB in calling for companies in hazardous chemical processing to establish shift work policy to minimize the risks associated with sleep-deprived workers.

1.3 A "WAKE-UP" CALL FOR THE PROCESSING INDUSTRY

The exceptionally demanding work schedules for the ISOM crew at BP Texas City are not uncommon. Workers can be driven to work long shifts without time off for overtime incentives. Management has economic pressures to return to production as quickly as possible. The CSB and others called on management to establish shift work policies with the goal of minimizing the effects of fatigue and reduce the hazards associated with sleep-deprived workers monitoring and operating dangerous chemical processes.

In response to this call for an industry guideline for fatigue risk management, the American Petroleum Institute (API) and the American National Standards Institute (ANSI) created the *Fatigue Prevention Guideline for the Refining and Petrochemical Industries*. This recommended practice (RP) provides guidance to management and workers on understanding, recognizing, and managing fatigue in the workplace. The guideline, ANSI/API RP 755, calls for companies to develop a fatigue risk management system (FRMS). These systems should include education on the biology of sleep and the potential dangers of fatigue and sleep disorders. The goal of the sleep education is to make workers, their families, and supervisors aware of the health reasons for sufficient sleep as well as the safety concerns. The API guideline also addresses limiting hours and the number of days operators can work.

The FRMS should be developed for the specific organization and be tied to other operational and safety procedures. The FRMS should address the following topics:

- Positions in a facility covered by the FRMS;
- Roles and responsibilities of those covered by the FRMS;
- Staff workload balance assessments;
- Safety promotion, training, education, and communication;
- Work environment;
- Individual risk assessment and mitigation;
- Incident/near miss investigations;
- Hours of service guidelines;
- Call-outs;
- Exception process; and
- Periodic review of the FRMS to achieve continuous improvement.

API RP 755 is only a handful of pages long and is a recommended practice rather than an enforceable regulation. Yet it is a clear indicator that the process industry has recognized the importance of human fatigue as it relates to safety. It highlights the importance of fatigue countermeasures. Unlike early efforts to limit duty hours, the FRMS also addresses the other factors that influence human performance and fatigue-related safety [4].

Sleep researchers continue to demonstrate the negative effects of sleep deprivation on workers. Decision-making skills decrease, both in monotonous and emergency situations, when workers do not have sufficient sleep [5]. Group interaction and decision-making are also affected by fatigue [6]. Researchers have shown that individuals who have gone without sleep or have an accumulated sleep debt perform comparably or worse than those with an elevated blood alcohol level [7].

COST OF SLEEP LOSS TO INDUSTRY
A recent study estimated the prevalence of insomnia was 23.2% and that it resulted in an average lost work performance equivalent to 11.3 days of work per individual. This was estimated for the total US workforce to be a $63.2 billion loss.

The costs associated with the consequences of sleepiness on the job can be astronomical (litigation, accidents, productivity, health care, etc.), and the impacts can be pervasive across families, communities, and organizations. It is important for managers and safety professionals to understand how sleep affects human physical and cognitive performance. This knowledge can be used to address possible risks and develop solutions. Although there is no simple, universal solution to fatigue in the workplace, a variety of countermeasure strategies have been proposed to maintain alertness and on-the-job performance. Something as simple as adjusting work schedules can have significant safety and performance benefits with a low cost. The design of an operator's workstation and work task may help reduce fatigue and improve performance. While the BP Texas City accident was not solely caused by human fatigue, it was a clearly a very significant factor. It is important that we understand human limitations and heed the "wake-up call" of API RP 755 and improve industrial safety. The processing industry must sincerely implement FRMSs and not wait for another tragedy.

REFERENCES

[1] CSB Texas City Final Report. Retrieved from http://www.csb.gov/assets/1/19/csbfin-alreportbp.pdf
[2] Orzeł-Gryglewska J. Consequences of sleep deprivation. Int J Occup Med Environ Health 2010;23(1):95–114.
[3] US Chemical Safety and Hazard Investigation Board, Investigation Report No. 2005-04-I-TX Refinery Explosion and Fire, March 2007.
[4] Lerman SE, Eskin E, Flower DJ, George EC, Gerson B, Hartenbaum N. American College of Occupational and Environmental Medicine Presidential Task Force on Fatigue Risk Management. Fatigue risk management in the workplace. J Occup Environ Med 2012;54(2):231–58.

[5] Harrison Y, Horne JA. The impact of sleep deprivation on decision making: a review. J Exp Psychol Appl 2000;6(3):236–49.

[6] Barnes CM, Hollenbeck JR. Sleep deprivation and decision-making teams: burning the midnight oil or playing with fire? Acad Manage Rev 2009;34(1):56–66.

[7] Williamson AM, Feyer A. Moderate sleep deprivation produces impairments in cognitive and motor performance equivalent to legally prescribed levels of alcohol intoxication. J Occup Environ Med 2000;57(10):649–55.

Basics of sleep biology

One of the cornerstones of any fatigue risk management system is an education program to communicate the principles of sleep and circadian biology. An understanding of basic sleep and circadian science will help employees and supervisors adapt these principles to the unique challenges that comprise each work environment. Not every scenario can be anticipated and have a predetermined response; therefore having knowledge of the underlying biology may help all parties come up with the best and safest solution. In this chapter, we will introduce the major biological principles that govern sleep and wake regulation and discuss the consequences of sleep deprivation. In Chapter 3, Circadian Rhythms and Sleep-Circadian Interactions, the concept and impact of circadian rhythms and their interaction with sleep debt will be addressed. Sleep and circadian rhythms interact to determine the timing and quality of sleep as well as the ability to vigilantly and accurately perform tasks.

Life is constantly changing. The implications of this obvious statement are substantial. Numerous technological advances have made life much more convenient, informative, and many would suggest more enjoyable. We have endless information and entertainment at our fingertips through both the internet and television. With streaming services, anyone's favorite shows are available at any time and place for an endless duration. Other technologies in the past century have impacted sleep as well. The invention of the electric light bulb has revolutionized the way we schedule our lives, making it possible to be active throughout the night. Air travel has made every corner of the world accessible within a day for business or pleasure, altering circadian rhythms as one travels across time zones. Technological innovations such as televisions, smartphones, computers, and tablets have crept into and are now integral to our lives. Unfortunately, our biology has not adapted to the suddenness of these changes and they may negatively impact sleep in unanticipated ways. One such consequence is that it has been hypothesized that the electric light bulb is responsible for lost sleep [1], which may have contributed to a decrease in sleep times over the last century. This decrease in sleep is affecting our health and performance metrics.

Many variables determine whether a person is fatigued at any given moment when performing their work duties. Time on task, the difficulty or how engaging a task is, and the time of day, all affect our performance on a task at any given moment. Despite these other factors, one of the major determinants of fatigue is whether one has obtained the quality and quantity of sleep they require. How rested one feels at the outset of the day or at the beginning of the shift will provide the baseline from which the other factors will start to have their effects [2]. The more rested a person feels at the outset of their shift, the more one can counteract the fatiguing effects of the other inputs as one will start at a more alert and rested level. People are at their best, most accurate, and most energetic after a restful and consolidated night's sleep. Conversely, people feel rundown, display a lack energy, and have a whole host of complaints when they do not get adequate sleep. As will be discussed in this chapter, inadequate sleep decreases a person's quality of life, results in more emotional responses to situations, and numerous health and cognitive performance problems [3]. All of these consequences influence how a workplace functions and ultimately productivity. Therefore, it becomes important to understand what regulates the onset and maintenance of sleep, the consequences of sleep deprivation and what can be done if one cannot sleep so that both management and employee are aware of how important adequate sleep is and what measures can be taken to counteract sleepiness. Attending to one's sleep can establish a low baseline for fatigue, which can prevent errors and make for a better work environment. The benefits of a well-rested workforce include fewer errors, more productivity, and a healthier workforce, which can increase the bottom line for a business.

2.1 WHAT IS SLEEP?

Sleep is a recognizable behavioral state. Though the person sleeping is inactive, current evidence supports the notion that sleep is not simply a resting state. Nearly as many calories are used during sleep as during waking [4]. During sleep, there is a stereotypical progression of the brain through sleep stages, suggesting a necessary pattern [5]. These concepts support the idea that there is an active process that helps restore the body during sleep. Under this hypothesis, daytime activities tax the system as memories are made, physical activity is carried out, and one negotiates daily activities. During the sleep period, a restorative process is carried out to prepare the body and mind for the next day's activities [6]. Thus, a prominent hypothesis about the function of sleep results from a molecular restorative hypothesis.

Despite this prominent hypothesis, it is as yet unclear what exactly sleep restores or the mechanism by which restoration is carried out. On the basis of how people feel after consolidated sleep compared to how they felt before they slept, it is clear that something changes with sleep. In addition to restoration, there are other benefits to sleep. There is some energy savings that occurs during sleep and sleep helps to consolidate waking activity to the daytime for humans. People have evolved to find a safe, hidden, or protected place in which to sleep at night when is not optimal for them to be active [7]. Thus, sleep can occupy time that might otherwise be disadvantageous to be interacting with the world, such as night when our vision is impaired due to darkness. There is a large and increasing research effort to understand the function of sleep, help mitigate the consequences of sleep deprivation, and design better work schedules on the basis of optimal function of the human body.

Sleep is not a unitary process. When a person falls asleep, in many parts of the brain the activity decreases, yet in some parts of the brain the activity actually increases [8]. Different stages of sleep have different brain activity signatures, suggesting that they may have different functions. Also, the brain goes through a sequential set of events during sleep. This progression through the specific set of events has led to sleep researchers to label sleep as an active process [5]. One idea suggested by the progression of the brain through sleep is that the brain needs to go through this specific set of events in a specific order to accomplish the goals of sleep, and the order of these events may be critical to sleep's function. Cognitive and physiological deficits are seen in people with fragmented sleep even though they obtain nearly equivalent sleep durations [9–12]. Moreover, the duration and quality of sleep is dependent upon the types of activity performed, in which increased brain activity will result in increased sleep intensity. This "use dependent" hypothesis suggests these cells then detect this usage and enhance the restorative effects of sleep locally to meet their increased need [13]. Moreover, current sleep drive is related to past sleep history [14,15]. If one is sleep deprived and then completes a full day's activities, they will have more sleep drive than a person that does not start the day sleep deprived. These factors all work in concert to determine how sleepy one is at a given time. In Chapter 3, Circadian Rhythms and Sleep-Circadian Interactions, we will discuss circadian input to alertness, which has a major impact on performance at any given time of day.

2.2 IDENTIFYING SLEEP

Though we may not know exactly what happens during sleep, we do know how to recognize sleep in both humans and animals. Most people are familiar the polysomnography technique, often referred to as "PSG." They may recognize the numerous wires attached to the head to record brain activity.

This technique was used to define sleep stages and helped researchers to determine what happens in parts of the brain during sleep. But sleep did exist prior to the ability to record brain waves using a defined set of behavioral characteristics that differentiate it from waking activities. Sleep has been defined in numerous animals, even in animals for which we cannot measure the brain waves. Sleep can be defined using several behavioral criteria [16]:

1. Quiescence: When an animal sleeps, it is still. Under normal circumstances, movement is typically indicative of goal directed activity, which is defined as wakefulness.
2. Increased arousal thresholds: Arousal thresholds are the minimum level of stimulation that results in person responding to a stimulus. An example of arousal threshold is when someone calls a person's name. If awake, one will typically respond by turning and addressing the person. While awake, even whispering the person's name will elicit a response. When one is asleep, the volume of the name would need to be louder to cause the person to respond. Thus it takes an increased stimulus to elicit the directed behavior. These arousal thresholds increase as one falls asleep and change throughout the night with different stages of sleep.
3. Rapid reversibility: The animal should be able to be aroused quickly from sleep. Though arousal thresholds change over the course of the night, there should be a level of stimulation that will wake the animal up. This criterion distinguishes sleep from other conditions, such as coma, anesthesia, or even death.
4. Homeostatic compensation: Though this sounds complicated, it is something we all can relate to. Simply, homeostatic compensation occurs with long periods without sleep. Sleep drive will continue to increase, and a sleep deprived person will sleep for a longer period than normal at the next sleep opportunity, a process also known as a "sleep rebound." This suggests that sleep plays a crucial role because we need to make up lost sleep. There are parallels to our eating response. If one misses a meal, there will be a compensatory increase in the amount of food that one eats but it likely will not constitute two full meals worth of food. In the same way with sleep, the body has an internal accounting mechanism that accounts for how much sleep has been missed. Humans and many other animals tend to regain one-third to one-half of the lost sleep. In the work culture, recovery sleep often occurs on weekends or other days off when people are allowed to sleep naturally without the timetables of work schedules. These individuals are fulfilling a homeostatic need for lost sleep.

5. Circadian rhythms: The last criterion is that sleep typically follows a circadian rhythm. In other words, sleep occurs at a particular time each day. This criterion is not absolute as we can willfully delay sleep, but our internal clock is programmed to promote sleep at particular times of the day. These criteria help define sleep for research and to adapt to novel situations in scheduling or difficulties sleeping.

In humans and other mammals, PSG can further subdivide sleep into the characteristic stages on the basis of electrical activity of the brain. Electrodes on the skull measure the neuronal activity from the surface of the brain, known as the cortex. This portion of the brain is a more contemporary portion of the brain and is responsible for some of the higher order intellectual processing. These are not recordings of single neuron, but are bulk recordings of neuronal activity from a small portion of the brain. Cortical recordings indicate that there are three distinct brain states, wakefulness (W), nonrapid eye movement (NREM), and rapid eye movement (REM) sleep. EEG measures the neuronal activity, but we will not understand each of the function of the stages of sleep until sleep scientists define molecular mechanisms that occur during each of these stages. Each of these stages of sleep is discussed briefly and note that these descriptions are group averages for healthy individuals [5] (see Fig. 2.1a for an example of each of the brain waves).

- Wakefulness: During wakefulness, PSG pattern exhibits low-amplitude, high-frequency waves. This type of pattern is thought to be generated by the asynchronous firing of the neurons during the sophisticated input of sensory information to the brain, the processing of that information, and the output to respond to the changing environment. Moreover, the brain is constantly changing to form memories. Because of all of this activity, the overall EEG during wakefulness exhibits high-frequency, low-amplitude voltage changes in cortical activity (Fig. 2.1).
- NREM sleep: NREM sleep is not a singular process and it progressively changes throughout the night. There are currently three stages of NREM sleep that have different EEG characteristics. Physiologic changes that occur during NREM sleep include a slow and regular breathing, a slowed heartbeat, a decrease in muscle movement (but it is still possible), lower body temperature and energy consumption, and a general increase in arousal thresholds with higher NREM sleep stages [5,17].
  - ❑ NREM Stage 1: Stage 1 represents a transitional state in which the brain is moving from wakefulness to sleep. Thus, the EEG takes on features of both the waking EEG as well as the beginnings of

(a) Behavior stage

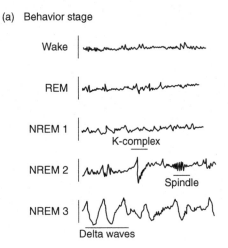

Wake

REM

NREM 1

K-complex

NREM 2

Spindle

NREM 3

Delta waves

(b)

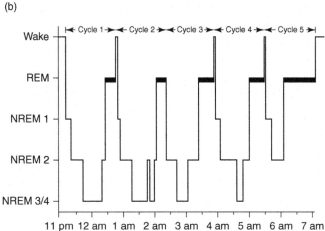

■ **FIGURE 2.1 Typical aspects of sleep in a healthy young adult.** (a) Representation of brain waves from EEG measurements at each stage of sleep and wake. As the brain goes into deeper stages of sleep of nonrapid eye movement (NREM) sleep, neuronal firing becomes more coordinated and the brain waves decrease their frequency and increase their amplitude. When the brain goes into rapid eye movement (REM) sleep, brain waves appear similar to wakefulness. (b) Representation of a hypnogram of a healthy person. Hypnograms describe the duration and temporal order of sleep stages. Thick bar denotes REM sleep. Note the descent through the NREM sleep stages into REM sleep, sleep cycles are ~90 min, and SWS is more prominent in the first half of the night and REM sleep more prevalent in the second half.

sleep-like oscillations in the recordings. Given that Stage 1 sleep is in this transition phase, a healthy sleeper does not spend much of their night's sleep in Stage 1. The person begins to display the slow, rolling eye movements associated with sleep. It is relatively easy to rouse a person from Stage 1 sleep and it does not appear to be restorative. Stage 1 sleep is transitory and only occupies 1–7% of the night's sleep. If a person spends too much time in Stage 1 sleep, it could be a hallmark of inefficient or abnormal sleep.

❑ NREM Stage 2: In the normal first sleep cycle, NREM Stage 2 sleep typically follows NREM Stage 1. In subsequent sleep cycles, NREM Stage 2 may follow other stages as well. With Stage 2 sleep, the PSG begins to show slower oscillation frequency and an increase in amplitude. Stage 2 sleep has two unique electrophysiological occurrences. The first is the K-complex, which is a large increase in the potential and then a sharp decrease in the potential recorded in the brain. The PSG trace quickly returns to baseline levels. The other unique feature is the sleep spindle which is a fast oscillation between higher and lower voltages with a gradual increase in followed by a decrease in amplitude. The K-complexes and the sleep spindles represent specific connectivity between brain regions that function during sleep that may be responsive to external stimuli during sleep [18,19]. This stage typically occupies 45–55% of the sleep time.

❑ NREM Stage 3 (slow wave sleep): In NREM Stage 3, the PSG trace begins to show the hallmark slow waves, which are portions of the EEG in which there are low-frequency, high-amplitude oscillations. Formerly, this stage was divided between Stage 3 and Stage 4 sleep based on the percentage of time in which the brain exhibits slow waves. For Stage 3 is was 20–50% whereas for Stage 4 it was greater than 50% of the recording time. Recently, these two stages have been combined as it is difficult and potentially arbitrary to discriminate between the two [20]. Slow waves reflect the coordinated and alternating firing pattern observed in cortical neurons. In other words, neurons will fire together in short bursts and then stop firing for a period only to begin burst firing again throughout the cortex. Slow waves are thought to be a very important feature of sleep and have been hypothesized to reflect the restorative capacity of sleep. In a normal individual, NREM Stage 3 occupies about 15% of sleep time (Fig. 2.1).

- Rapid eye movement (REM) sleep: The name, REM sleep, comes from the hallmark eye movements that can be seen underneath the closed eyelids during this stage of sleep. This stage of sleep might be the most recognizable and fascinating stage of sleep for the average person. REM sleep is a critical component of sleep as REM deprivation can result in the death of animals [21], similar to total sleep deprivation. Although all stages of sleep have been shown to have an impact on learning and memory, evidence suggests that REM sleep may have a particularly important role in memory [22] as well as replay some of the day's events that increase procedural memory and creativity [23]. REM sleep usually occupies 20–25% of an individual's sleep time and may have subdivisions because the eyes are not always moving while the brain appears in REM sleep, though they are not yet considered significant enough for official subsets [20]. The first occurrence of REM sleep usually occurs after one has progressed through each of the NREM sleep stages listed previously. Initially, REM sleep was called "Paradoxical Sleep" because the brain waves of the individual had a high frequency and a low amplitude, a signature very similar to the waking brain yet behaviorally the individual appeared asleep. An awake brain showed this type of pattern because it was processing and integrating information, but the unexplained paradox was why the asleep brain would exhibit an asynchronous PSG.

The answer lies in the fact that REM sleep is probably best known for its association with dreams, especially the memorable, emotionally charged, and fantastical dreams [24]. In fact, a fascinating thing occurs during REM sleep, our brain may disengage the part of the brain that applies the rules of what is real to our sensory input. This frees our brain to conjure images and associate things that normally would not be permitted to be associated with one another. Therefore, this stage of sleep has been looked on as partly responsible for some of our creative solutions to difficult problems. In fact, one of the most famous examples is to have come from chemistry. It is rumored that the German chemist, Augusta Kekulé, may have come up with the structure of benzene during a dream. The scientist could not deduce the structure until one night during he dreamed about a snake biting its tail. This led to the insight that benzene is a ring structure. Though this particular story may be more myth than truth, it highlights a point that sleep is associated with increased creativity and insight [25]. This free association may be an integral part of this process. But while we dream, our bodies carry out another unique aspect of REM sleep. We are unable to act out our dreams because our movement is paralyzed during REM sleep. Through a neurological process, all of our movements except for our eye movements and breathing are blocked. When this process breaks down, it may be a clinical problem known as REM behavior disorder (see Chapter 5, Sleep Disorders). These patients can harm themselves and their bed partners while acting out their dreams. Thus REM paralysis serves an important function to protect us during sleep.

REM sleep also has very different physiological characteristics than NREM sleep. First, heart rate increases and becomes erratic. In conjunction, blood pressure increases and becomes more variable. Breathing rate also becomes more erratic and deeper than that observed with NREM sleep. REM sleep can be accompanied by sexual arousal. Arousal thresholds during REM sleep are at times elevated and at other points decreased [5]. These changes are used to help determine whether an individual is in NREM or REM sleep.

The control of sleep is a biochemical process. Neurotransmitters are the workhorse of sleep regulation. Different combinations of neurotransmitters are associated with different stages of sleep. This combinatorial code can be disrupted by age, disease, or pharmaceuticals [26]. One common example is common prescription sleeping aids enhance the effect of the neurotransmitter γ-aminobutyric acid (GABA), which helps decrease the activity of the brain and promote sleep. Age and disease may also alter the balance of these biochemicals in such a way that sleep is also altered. These changes may be the cause of or result from sleep disorders, which will be more fully discussed in Chapter 5, sleep disorders. In addition to neurotransmitters, inputs from molecules including metabolic by-products [27], whether we have eaten recently or not [28], and immunological molecules can regulate sleep [29]. Sleep regulation is a complex process with numerous controls that favor sleep or wakefulness. Therefore, it can be manipulated purposefully or inadvertently through pharmaceuticals or from other external or internal factors.

To visualize how sleep progresses through the night, a graphic called the hypnogram was created (Fig. 2.1b). A hypnogram displays the duration that an individual stays in a given sleep stage over the night. This graph can be used to determine the percentage of time that an individual spends in a given sleep stage for the night as well as the relationship between the sleep stages for that particular person. Under normal circumstances, a person enters sleep through NREM sleep and the stages of sleep progress through NREM Stages 1 through to Stage 3 and then into REM. In an adult, this process takes roughly 90 min to complete a full sleep cycle. Thus, one will achieve 4–6 sleep cycles over the night. The average time for a complete cycle becomes important as one tries to plan out naps and if one wakes up before sleep debt is fully dissipated. When one wakes from a deeper stage of sleep, it can result in performance decrements initially after waking (see Sleep inertia below and in Naps, Chapter 14). Moreover, the distribution of sleep stages is not the same over the night. In the early part of the night, a greater portion of sleep is dedicated to slow-wave sleep (Stage 3 sleep) than is dedicated to REM sleep. As the night progresses, a greater percentage of sleep time is dedicated to REM sleep compared to slow-wave sleep. This is why it seems that one is having a dream while the alarm clock is going off or why the alarm clock seems to get incorporated into one's dreams.

Sleep deprivation alters the distribution of sleep. After a night of sleep deprivation, recovery of slow wave sleep is prioritized over other stages of sleep, even though all stages of sleep are lost [30]. Thus, both the normal sleep cycle data and the sleep deprivation data would suggest slow wave sleep is a particularly important stage in the restorative function of sleep. Though slow wave sleep is important, it appears that all stages of sleep contribute to optimal performance and health [22,31–33]. Also, depriving an animal of any stage of sleep will impact cognitive performance. Since sleep cycles are a critical part of sleep, it is reasonable to hypothesize that the stages interact and carry out a particular molecular function that carries out the restorative function during sleep.

2.2.1 **Sleep fragmentation**

To fully carry out the restorative function of sleep, the sleep needs to be both of sufficient duration and consolidation. Insufficient sleep duration will result in a sleep-deprived state because the body is unable to dissipate the sleep debt that has built up over the day. Just as importantly, sleep cannot be interrupted by wakefulness too often, also referred to as sleep fragmentation. With each interruption, the sleep cycle is more likely to revert back to stage 1 and start the sleep cycle over again. Thus, a person with fragmented sleep may not spend an adequate amount of time in deeper stages of sleep. Even without reducing the total sleep time, sleep fragmentation can result in health and cognitive problems [10–12,34]. Sleep fragmentation results in increased sleepiness, reduced cognitive performance, and glucose processing problems. These are similar effects to those seen with sleep deprivation. Sleep fragmentation increases with age [35]. Also, sleep fragmentation is seen in many sleep disorders, such as sleep apnea [9]. These data support the conclusion that both sufficient sleep duration and consolidation are important for optimal health.

2.2.2 **Sleep inertia**

Another property of sleep that has been recently recognized is referred to as sleep inertia. Inertia is the concept that a body in motion will continue to be in motion whereas a body at rest will continue to stay at rest. As it applies to sleep, sleep inertia is a difficulty of the brain transitioning from the sleeping state to the fully awake state. In the period immediately following sleep, is a person may have a harder time feeling as though they are at full cognitive capacity or having "cobwebs." This period of sleep inertia results in lower cognitive performance and potentially poorer performance than being sleep deprived [36]. This concept has implications for scheduling rest breaks, naps, and shifts. The impairments of sleep inertia can last up to 1 h or more [37]. If one wants to ensure that an employee is at optimal performance, the idea of sleep inertia needs to be accounted for employees that are able to nap during work or employees that are on call. For naps, it is recommended to take a short nap, typically about 20 min, so that one gets the restorative effects of sleep without getting to deeper stages of sleep, which might increase sleep inertia. Employees on call should be given a period to wake up and others communicating with them should understand their counterpart's decision-making situation. The effects of sleep inertia are demonstrated by military pilots that have a greater likelihood of crashing within 2 h of sleeping. It was concluded that the pilots needed some time to adjust from sleep to the quick decisions that needed to be made when flying a plane [38]. It is likely that transitioning from a state that is coordinated in a much different way, such as slow wave sleep, to complete wakefulness is not a unitary and immediate process. Sleep inertia may make it difficult to have the cognitive capacity to integrate information and make critical decisions and formulate the appropriate response. Thus, sleep inertia is a concept that needs to be factored into decisions about scheduling and worker performance.

2.3 **WHAT IS SLEEP GOOD FOR?**

Sleep is a requirement for every animal that has thus far been rigorously studied. When sleep deprived, there is a proportional increase in sleep, called "recovery sleep," that makes up for the lost sleep. In addition, multiple types of animals die when they are deprived of sleep in the long term [39,40]. These results all imply that sleep is carrying out some critical function. Scientists are unclear about what that function is precisely. A leading hypothesis is that sleep carries out a restorative function and allows the cells be prepared for the coming day's activities. At the cellular level, this restoration would repair and replace used proteins to permit the efficient functioning of the cell. Another hypothesis suggests a more functional role for sleep which is to consolidate memories and/or prepare the brain learn again the next day. These two hypotheses are not mutually exclusive. More practically, a lack of sleep results in numerous health and cognitive consequences that occur with sleep deprivation, which may provide a clue for why we need to sleep. Each of these hypotheses has solid evidence that supports it. Given the amount of time that we sleep, it is possible that sleep does not have one particular or primary function and that each of these hypothesized functions are carried out during sleep.

2.3.1 **The concept of Process S**

Despite the lack of evidence for what occurs at the molecular level, sleep scientists have been able to elucidate a lot of information for what occurs at a behavioral and conceptual level during sleep. The fundamental principle can be summarized as the longer one is awake, the stronger the drive to go to sleep is. This can be encapsulated in the idea of "sleep debt." Evidence suggests that there is some sort of system that tabulates how long someone has been awake and how much activity that person has carried out. This accounting manifests as an increase in sleep pressure that will result in one falling asleep. This idea was formalized in the two process model of sleep regulation by Borbely and colleagues in the early 1980s. Under this proposal, as waking continues, sleep debt continues to accumulate and is dissipated with sleep [41,42]. In addition, a second system, the circadian system governs the timing of sleep and alters cognitive performance on the basis of optimal biological times for sleep. Under normal circumstances, this model demonstrates why we are the freshest in the morning, why we are the most tired at night, at what rate sleep debt builds up, and how quickly it is dissipated (see Chapter 3, Circadian Rhythms and Sleep-Circadian Interactions). The accumulation of sleep debt is summarized as Process S (Fig. 2.2a). The model also begins to predict how past sleep history influences the current sleepiness of an individual (Fig. 2.2b) and has been successfully applied to

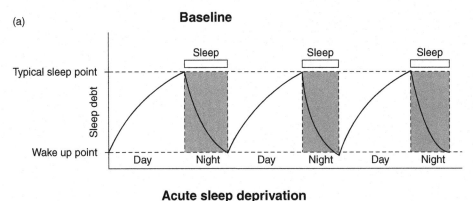

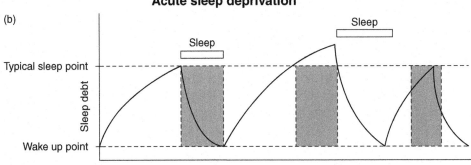

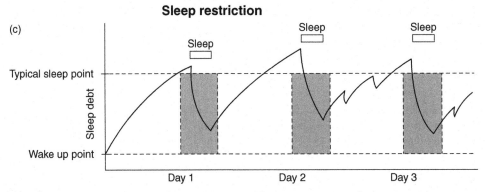

■ **FIGURE 2.2 Patterns of sleep debt with different types of sleep deprivation.** Sleep debt reflects the likelihood to fall asleep and has a complex relationship with cognitive performance. This figure reflects only sleep debt without the influence of circadian rhythms, which this additional complexity will be added in the next chapter. (a) Under a routine schedule, sleep debt builds up during wakefulness (typically during the day) until it reaches the point that it induces sleep at the end of the wakeful period. Sleep debt is dissipated with sleep (typically at night). This pattern repeats itself each day, especially when one has a regular bedtime. Note that neither buildup nor dissipation is linear. (b) Individuals can willfully put off sleep. Under acute sleep deprivation conditions, in which one misses sleep for a night, sleep debt builds up past when sleep is typically induced. Though one can remain awake, the pressure to go to sleep builds up and puts even more sleep pressure on the individual. Thus, performance during extended wakefulness will likely be degraded by sleepiness. (c) Sleep restriction is obtaining less than adequate sleep each night. Over successive nights, sleep pressure reaches higher levels with repeated inadequate sleep. In addition, data has postulated that there is a ceiling to the amount of sleep debt accumulated with sleep restriction. One hypothesis is that during the waking period some of the sleep, debt is dissipated through processes such as microsleeps. Thus during the waking period, sleep-like processes intrude into wakefulness causing lapses and inattentiveness in people and they may not be able to tolerate additional sleep deprivation without showing additional cognitive problems. Another hypothesis suggests that sleep debt dissipates faster under sleep restriction conditions.

predict what the best work schedule that would optimize work performance based on sleep and circadian factors [43]. These models and concepts have continued to progress to more accurately describe the performance of individuals under an increasing set of conditions, including incorporating sleep inertia and longer periods of sleep and longer periods of sleep restriction (Fig 2.2c), both of which add complexity to the systems.

2.3.2 How much sleep does one need?

First, let us be clear, there is no given answer for a specific individual. The answer will be based on the needs of that particular individual. There are people whom sleep researchers call "short sleepers" who require reduced amounts of sleep and there are those called "long sleepers" who require higher amounts of sleep. These individual differences may be due to the way that one's sleep is inherently structured. Unfortunately, many people believe that they are short sleepers, even though there are very few people that have this little sleep need. Indicators that one is not obtaining the adequate amount of sleep are that it takes an alarm clock to waken one at the appropriate time, one will sleep in on weekends or on a day off, and that one experiences excessive daytime sleepiness that interferes with alertness and performance. Ultimately, the goal of sleep needs to be fulfilled before sleep can end. The goal of sleep is to dissipate the sleep debt through the underlying molecular mechanisms. Recently, the scientific evidence was evaluated how much sleep was optimal for people. They concluded that as a population between 7 and 8 h yielded the best health and performance outcomes in adults [44]. But as stated previously, each individual has their own sleep need and it may change on the basis of sleep history and other factors.

Sleep changes over a lifetime, so depending on how old a person is the average amount of sleep time is different:

Infant/childhood: As infants, sleep time is at its longest. Upon birth, sleep is structured differently than it is as an adult. There are periods of active sleep and quiet sleep, which may reflect an immature version of the typical NREM and REM observed in older childhood and through adult stages. Also, in the first months after birth, infants have not yet fully developed circadian rhythms but they develop by between 3 and 6 months of age [45]. Therefore, sleep is often distributed evenly over the day and night, which means that new parents often will have fragmented sleep while tending to the needs of the newborn. With time, the structure of sleep develops into the adult pattern. On average, sleep in infants averages 14 h per day but individuals can range from 9 to 19 h [35]. As infants develop into children, average sleep durations decrease steadily, but even by age 16 average sleep is still 9 h per day, longer than observed for adults. Sleep is important for a developing child, as inadequate sleep for children can have consequences on the cognitive development [45]. Because sleep in infants can be spread throughout the day and night, tending to a small child can disrupt and fragment the parents' sleep.

One important point is that a child's sleep duration and timing of sleep needs to be based on the individual child. There are numerous factors that go into the amount of sleep that a child actually gets. First, there are the genetics factors, which control how long a child will sleep under ideal circumstances. This timing reflects the buildup of sleep debt during waking periods and the dissipation of sleep debt during sleep. It also reflects the mechanisms to initiate and maintain sleep, which can be a product of both sleep regulatory mechanisms as well as circadian changes that occur throughout childhood development. As a child ages into adolescence, the timing of sleep typically drifts farther into night. Therefore, a teenager may not be able to fall asleep at a time that would permit a full night's sleep. Moreover, some children are just "night owls" because their circadian cycle is delayed. It is difficult to put these children down early because their body is promoting wakefulness at that time of night instead of permitting sleep. Second, sleep durations reflect the culture of both the parents and the age of the child. The lifestyle and sleep habits of the parents can influence whether the child in permitted to get enough sleep. In addition, as the child ages school responsibilities and social opportunities begin to interfere with adequate sleep. It is recommended that the parents work with the biology of the child to promote adequate sleep time. Parents should be aware of the rhythms and the sleep needs of their child and use these tendencies to help promote adequate sleep in their children. One metric to know if one is successful is to determine whether the child is using "catchup" sleep on the weekends to dissipate sleep debt when there are no responsibilities forcing wakefulness at a particular time. If so, then it is likely that the child is not getting sufficient sleep during the week. Hopefully, school systems will begin to take into account the mounting research suggesting that students are not getting sufficient sleep during the week and that as children age their sleep need and patterns change. Accommodations for these changes can result in better academic performance and better behavior in school. *Adults*: Experimental evidence is attempting to determine the optimal sleep time for adults. Several different lines of evidence suggest that sleeping between 7–8 h is what is required, and a new report from numerous sleep scientists supports this conclusion [44]. Epidemiologic data has demonstrated that the healthiest weight people sleep around 7.5 h [46]. Sleep durations above and below this duration are associated with increased BMI. In addition, all-cause mortality is lowest at 7.5 h of sleep [31–33]. Deviations from this duration higher or lower results in increased mortality rate. Thus, the data coalesce around approximately 7.5 h for adults as a population average. *Elderly*: By the time one reaches age 60, there are numerous changes that occur the sleep characteristics. From population studies, there are decreases in total sleep duration, NREM Stage 3, and REM sleep. There are increases in NREM Stage 1, sleep onset latency, and wake after sleep onset. These data suggest that sleep quality and quantity decline with increased age. It is unclear that sleep need in the elderly decreases. It has been hypothesized that sleep need does not change but that the ability to obtain consolidate sleep of the proper duration is compromised. In healthy aged adults, there is less of a decline in slow wave sleep. The difference between those that maintain sleep quality and those that do not could be due to increasing health problems that accrue as one ages [47]. In survey results, elderly adults with more diagnosed health problems, such as cardiovascular, high blood pressure, or sleep disorders, more frequently complained of sleep problems. The increase in the sleep disruption arises from two different sources. First, the health problem itself may be causing the sleep disruption. Each malady may increase pain or some other discomfort that will interfere with sleep. Second, the treatments for these conditions may disrupt sleep. Both over the counter and prescription drugs can alter the biochemistry of sleep regulatory system. Thus as the number of drugs and maladies increases, the likelihood increases that sleep will be disturbed in some fashion. These findings have led some to propose that absent any health problems, sleep in the elderly should not necessarily decline more than a modest amount. At the same time, advancing age is a risk factor for nearly every major health problem, so sleep problems in the elderly are common.

There is evidence that another outcome of aging is that circadian rhythms begin to "flatten." This is defined as the peaks of the circadian rhythms are not as high and the lows of the circadian rhythms are not as low [48]. Under these circumstances, night and day do not have as robust biological differences as when the person was younger and the rhythms for going to bed and getting up as well as maintaining sleep during the day are all not as strong and well defined. Naps can begin to creep into the daytime and wakefulness begins to creep into the nighttime period. Although napping in the elderly is associated with negative health outcomes in the long run, there are beneficial changes to performance and alertness [49]. Just as sleep may not necessarily decrease in healthy elderly people, circadian rhythms remain robust in healthy aged adults [50].

As we age, health and physiologic changes occur naturally that may impair our ability to sleep. These changes may apply to the company employee or a loved one that an employee may be caring for. Thus, identifying how the body ages and what a person can do to optimize sleep will enhance both health and cognitive performance or help understand what a loved one is going through.

2.4   CONSEQUENCES OF SLEEP DEPRIVATION

When it comes to interpreting sleep deprivation data, there are three types of data sets that may be providing different types of information about our sleep biology. The first is epidemiological data. In this approach, researchers evaluate a cross-section of the population and correlate it to given outcomes or characteristics. These types of data are often correlational, but they begin associations that can be tested in other ways. The other two approaches are experiment-based approaches in which people are tested under baseline conditions and after a sleep deprivation protocol. The second approach is acute sleep deprivation, in which a person is kept up for an entire night or longer. Although this is an effective strategy for research, under real-world conditions people are rarely up all night. To address this concern, a newer sleep deprivation protocol is called sleep restriction (Fig. 2.2c). These protocols mimic a more common sleeping experience by restricting sleep to a shorter duration, such as 4 or 5 h a night for about 5–14 days. This is a more time-consuming and expensive approach but it has yielded some interesting results that differ from acute sleep deprivation. Additionally, there are more complicated protocols that control for circadian rhythms which we will not deal with here, Chapter 3, Circadian Rhythms and Sleep-Circadian Interactions. Sleep experts are now aggregating data from epidemiological, acute-sleep-deprivation, and sleep-restriction experiments to understand what happens to people when they do not fully attend to their sleep. As will be shown, the consequences of sleep deprivation have an impact on the on-site performance of the employee, the productivity of the employee, and health-insurance costs.

2.4.1   Cognitive performance

2.4.1.1   Sleepiness

Sleep scientists have documented what most people intuitively understand about the most conspicuous effects of sleep deprivation, there is a substantial increase in the drive to sleep. Although this may seem an obvious outcome for a sleep-deprived person, this observation has important implications on how and what we do. This can lead to someone falling asleep when they get to a sedentary, comfortable, and monotonous situation, such as driving. A recent survey shows that 60% of poll respondents have driven while drowsy in the last year, 37% had nodded off or fallen asleep while driving and 4% admitted to having an accident or a near miss because of drowsiness [51]. A conservative estimate puts the number of sleepiness related accidents at 100,000 per year. Studies of residents on extended shifts showed a significant increase in automobile accidents, near misses, and falling asleep at the wheel after extended shifts [52,53]. This same applies to truck drivers, pilots, and others that have critical job in the transportation, healthcare, and safety monitoring or maintenance industries in which people's lives depend on proper job performance.

One fact that is clear is that people do not have a good sense of how sleepy they are and/or they do not listen to their body telling them they are close to sleep. Therefore, increasing sleep deprivation or sleep restriction will continue to increase the sleep debt. With increasing sleep debt, there is an increasing likelihood that one will fall asleep, potentially at a dangerous or inopportune moment. Also, with increasing sleep deprivation, one is more prone to microsleeps during wakefulness. Microsleeps are involuntary lapses in attention that can erode performance [54,55], which are not as obvious as when one falls asleep but can be just as devastating. During the period of a microsleep, the person is awake with their eyes open but they do not respond to incoming stimuli even though they are expecting to have to respond to the stimuli. Also during these periods, the brain exhibits transient high-amplitude, low-frequency waves that are the signature features of slow waves but during a waking period. In addition, when a person is sleep deprived, the EEG profile becomes more synchronized [56], suggesting changes in brain connectivity and processing with sleep deprivation.

2.4.1.2 *Cognitive performance*

Vigilant attention was one of the first recognized and most sensitive cognitive effects of sleep deprivation. The loss of vigilant attention can be investigated using the psychomotor vigilance test (PVT). In this test, the subject has a box with a button and a light. During a 10-min testing period, the light flashes at random intervals and the person responds by pushing the button as quickly as one can. There are many metrics that can be used to understand the vigilance state of the individual, but the most common are the response time of the individual and the number of times that a person takes more the 0.5 s to respond, referred to as lapses [57]. People that have been sleep deprived for 1 night exhibit longer reaction times and increased numbers of lapses. One impact of the decrement in vigilant attention is in learning.

In order to learn something, the person must first be able to attend to it. Thus, one's ability to recall novel information is impaired with sleep deprivation.

This phenomenon does not just occur after just 1 night of acute sleep deprivation. Disrupted vigilance also occurs after repeated nights of sleep restriction. In a laboratory study in which people were restricted to 0, 4, 6, or 8 h of time in bed for 14 days, increasing sleep times were positively correlated with better performance on the PVT. Even when people obtained 6 h of sleep per night, their cognitive performance on the PVT continued to decline over the sleep restriction period [58]. Within a week, the 6 h time-in-bed group performance declined to the point equal to a full night's sleep loss and eventually exhibited performance equivalent to 48 h of continuous wakefulness. Interestingly after several days of sleep restriction, these subjects did not report feeling more tired but their cognitive performance continued to decline. This test reinforces the points that people are not good judges of their own sleepiness and that skipping even 2 h of sleep per night has significant cognitive consequences.

The sensitivity of higher order cognition to sleep deprivation has produced mixed results. Tasks such as a digit substitution and computing mathematical problems are also very sensitive sleep deprivation [59]. As the tasks become more complex and engage more aspects of planning and creativity, there appears that cognitive engagement that is not as sensitive to sleep deprivation or that it is difficult to measure which aspects are sensitive [60]. Recent data suggests that performance on these types of tasks may decline due to sleep deprivation. One higher order task that is altered is risk taking. Under sleep deprived conditions, individuals may be more likely to take risks [61]. Therefore, sleep deprivation alters how we process and respond to complex input.

2.4.1.3 *Emotion, vigor, quality of life*

Inadequate sleep alters the emotional aspects as well as cognitive decrements. Sleep deprivation has been associated with a change in how we process emotions. One example is that comes from experiments demonstrating that sleep deprived individuals may misinterpret facial cues and attribute more emotional importance under sleep deprived conditions than under well-rested conditions [62,63]. This may pave the way to misunderstandings in the workplace or in dealing with colleagues. In fact, sleep deprivation results in a decline in how parents relate to one another [64]. Moreover, inadequate sleep is associated with increased levels of depression [65]. In addition, decreased sleep times are associated with decreased vigor and decreased outlook [66]. These impair one's ability to enjoy life and can take some of the pleasure out of life. They can also impair how coworkers interact with one another or affect how a person attends to their job.

2.4.1.4 *Presenteeism, tardiness, and absenteeism*

Inadequate sleep is correlated with absenteeism, tardiness, and presentee-ism [67]. These three parameters cost employers billions of dollars in lost productivity. Individuals who have low-quality sleep, including those with insomnia, have increased days absent compared to employees with higher-quality sleep [68]. In an evaluation of multiple studies over a 25-year period, people who had difficulty sleeping would miss between 1 and more than 5 days more than their well-rested counterparts [68]. In addition, tardiness is also positively correlated with inadequate sleep. Although these two attri-butes are easy to distinguish, the newer problem of presenteeism is harder to identify. Presenteeism is defined as being on the job but not operating with optimal attention to work. Inadequate sleep leads to an increase in presen-teeism in self-reported data [3] and demonstrated a productivity decrease in employees obtaining inadequate sleep. Thus, the better that one sleeps, the less likely that one is to be sick or absent from work, which can increase the productivity for a business.

When it comes to cognitive consequences of sleep deprivation, each indi-vidual responds differently. People have been divided into sensitive, insensi-tive, and a middle group [69], but an individual's group was not necessarily consistent between cognitive tests. People who are insensitive to sleep de-privation have the same response on a task as they did under well-rested or baseline conditions. In contrast, sensitive individuals show numerous cognitive deficits after sleep deprivation. A person's genetics contributes to individual variability. Genetic makeup plays a role in a person's response to sleep deprivation in humans [70–72], and other genes will likely be identi-fied as there are numerous others found in other species [73]. Thus each individual will have a different response to sleep deprivation, which means thought must put into a general sleep-deprivation policy. What remains a mystery is the level of sleepiness in each individual. This will require ei-ther a physiological or biochemical biomarker, which is an active area of research.

2.4.2 **Health consequences of sleep deprivation**

Epidemiological and experimental data have uncovered numerous adverse health effects attributed to sleep deprivation. From an epidemiological per-spective, shorter sleep times have been associated with an earlier death [31–33]. The deaths may come from a health-related cause or from an ac-cident related cause, but obtaining an optimal amount of sleep leads to a longer lifespan in general. Sleep deprivation seems to affect a number of different processes, and more seem to be discovered every year. Health problems negatively impact both worker's productivity and health-related expenditures for businesses and employees. Some examples are:

2.4.2.1 *Cardiovascular*

There is increasing evidence that sleep is intertwined with cardiovascular health [74,75]. Given the impact of cardiovascular health on longevity, this is an important relationship for employers and employees. Epidemiological evidence suggests that sleep time is inversely correlated with blood pressure. In an evaluation of more than 5500 individuals, the group that obtained between 7 and 8 h sleep had significantly lower blood pressure than groups that either got more or less sleep than 7–8 h [76,77]. Moreover researchers carried out a longitudinal study in which they monitored changes in blood pressure in the same people after 5 years. They found a larger increase in blood pressure in people that had shorter sleep duration than people with a longer sleep duration. This puts increased stress on the entire cardiovascular system.

In addition, sleep deprivation is correlated with other cardiovascular effects. Sleep deprivation is correlated with increased heart rate over a 24-h period. This effect may result from an increase in the levels of the biochemical compound, norepinephrine, which has increased levels with sleep deprivation. Elevated level of norepinephrine with waking would increase both blood pressure and heart rate [78]. In addition, sleep deprivation is also associated with changes in the blood vessels that may promote hypertension [79] and increased arterial calcification [80]. These data, along with many other examples indicate that sleep duration impacts a person's both short-term and long-term cardiovascular health.

The connection between cardiovascular effects and adequate sleep is reinforced when people with sleep apnea, a sleep disorder that reduces oxygen intake and fragments sleep, are evaluated. Patients with sleep apnea exhibit deleterious cardiovascular changes, including elevated blood pressure [81]. These symptoms are reversed with the use of a device that helps one breathe more normally throughout the night [81]. These symptoms result, in part, from the sleep fragmentation that occurs with sleep apnea. Therefore, there is mounting data to suggest that cardiovascular health is diminished by inadequate sleep.

2.4.2.2 *Obesity*

One of the greatest worldwide health epidemics is obesity and associated metabolic disorders. Recent evidence suggests that inadequate sleep plays a role in this epidemic. In a cross-sectional study, the investigators observed a U-shaped curve relating sleep times to increased body mass index (BMI) [46]. People who slept about 7.5 h had the lowest BMI across the spectrum of sleep durations. With progressively lower sleep durations, the average BMI progressively increased. Interestingly, sleep durations longer than 7.5 h also showed elevated BMI.

Experimental evidence is beginning to identify mechanisms that may underlie this relationship. Levels of several endogenous chemicals that control food intake have been altered to favor food intake when people are restricted to 4 h of sleep for 5 nights [82,83]. These include peripheral hormones and neurotransmitters in the brain. These molecular changes may underlie the reason that people tend to consume more calories under sleep-deprived conditions [84,85]. The increased feeding tends to take place outside of the normal meal times, including snacks and eating after the last meal. Not only do the subjects eat more calories but the makeup of the calories also shifts. Subjects tend to consume more snacks and calorically dense foods, such as fats and carbohydrates [83–85]. In addition, sleep deprivation changes where energy comes from when one is attempting to lose weight. Sleep-deprived people will lose the same amount of weight as well-rested people but sleep-deprived people lose more lean weight as opposed to losing fat, which is the weight people want to lose [86]. Thus, there is mounting evidence that sleep deprivation has an impact on BMI and metabolism.

2.4.2.3 *Endocrine effect*

There are numerous changes in hormones that accompany sleep deprivation. The levels of two hormones critical to regulating feeding are altered in sleep-deprived people. The level of leptin have been shown to decrease while ghrelin increases [82]. Leptin signals satiety and ghrelin signal hunger, consistent increased hunger, decreased satiety, and eventually consumption of more food, though this may not be a universal mechanism [84,87]. In addition, levels of growth hormone are reduced [88]. Growth hormone is a critical for bone and muscle growth in children and adolescents as well as building and repairing muscles in adults. Growth hormone is released during slow wave sleep. Therefore, when one misses sleep, one will be missing the major release of growth hormone.

2.4.2.4 *Blood glucose levels*

Another prominent disease trend throughout the world is the increasing prevalence of Type II diabetes, or adult onset diabetes. In Type II diabetes, critical tissues like adipose tissue, muscles, and the liver lose their sensitivity to insulin signaling. This results in an increased blood glucose level after meals and results in numerous types of impairments. The mechanisms responsible for Type II diabetes are not well understood. Over the past decade and a half, an intriguing link has been established between sleep deprivation and impaired glucose uptake [89]. After several nights of sleep restriction, subjects exhibited an increase in plasma glucose after a meal compared to normal, well-rested controls. This prediabetic state was induced by a reduction in how the cells responded to the insulin thus removing less glucose from the blood less quickly. There was improvement in these subjects after 4 days of recovery sleep [82]. In a model organism, sleep deprivation disrupted insulin signaling dynamics within cells of adipose tissue [90], suggesting that sleep deprivation affects the molecules that process glucose and respond to insulin.

2.4.2.5   *Immune system*

Sleep and the immune system influence each other in a complex relationship [91]. Infection leads to an acute response in which sleep is increased and then typically a second phase in which sleep is more disrupted. On the other hand, sleep deprivation begins by increasing the levels of some immunological molecules [92,93]. In the longer term, sleep deprivation results in a less effective immune system [94]. Thus, routinely sleep-deprived people are more often sick and miss more work because of sleep deprivation [68]. Also, sleep-deprived individuals to not respond as well to vaccinations as do well-rested individuals [95] and have a propensity to get more colds [96].

Interestingly the signaling molecules in the immune system appear to significantly regulate sleep as well. Recently, a signaling pathway involving cytokines has been proposed to regulate sleep [97], thus when sleep deprivation raises the levels of cytokines it may impact the sleep regulatory pathway [93,98]. Thus, these immune molecules in the brain may mediate the increase in sleep observed after sleep deprivation. This cross-talk between these immune and sleep regulatory systems may explain why people will often get sleepy once they are sick and why sleepy people often get sick [99].

2.5   **BENEFITS OF SLEEP**

Much of what we believe we know about sleep comes from sleep-deprivation experiments. Under these circumstances, the scientist is investigating what happens if we take away sleep. This type of experiment will tell you what happens in the absence of sleep, but it does not tell what happens with sleep. There have been a couple of experimental protocols that appear to elucidate how sleep might benefit the person.

The first set of experiments uses a nap protocol. With the introduction of sleep, there were improvements in various cognitive assays. Even a 10–20-min nap is restorative enough to improve performance even though subjects may not feel more rested than controls that that did not take a nap [100]. Therefore, naps show that sleep is beneficial.

Another set of experiments attempts to induce sleep to test how sleep might improve various performance metrics. The manipulation of sleep was used to understand how sleep affects memory consolidation. In experiments that are very difficult in any other experimental system, one protocol used genetics to induce sleep in the fly after a training protocol that normally did not induce a memory in the fly. The induction of sleep improved memory consolidation and two days later the flies still remembered the task even though under normal circumstances the flies would not have remembered the training [101]. What makes this example unique is that this is the first instance of being able to genetically put an animal to sleep and demonstrate an improvement on a psychological parameter. In a complementary set of experiments, flies were trained on a task that they typically would remember 2 days later. They then were either sleep deprived immediately after training or permitted to sleep. Those who were permitted to sleep exhibited the memory 2 days later but those flies that were sleep deprived did not retain the memory. This phenomenon has been shown in humans as well. With a night's sleep, people are able to perform better on many cognitive tasks, including word recall. These experiments demonstrate that sleep benefits the cognitive performance of an individual.

Sleep is a complex process that impacts nearly every part of a human's physiology and cognitive aspects. In normal, healthy people, it has a stereotypical progression throughout the night and changes over both the night as well as over a lifetime. Despite its importance, inadequate sleep is prevalent in America today. There are consequences on both the cognitive performance and physiological parameters. On the basis of scientific data, sleep needs to be considered as one of the pillars of health, along with diet and exercise. Sleep needs to be prioritized by both employer and employees.

REFERENCES

[1] de la Iglesia HO, et al. Access to electric light is associated with shorter sleep duration in a traditionally hunter-gatherer community. J Biol Rhythms 2015;30(4):342–50.

[2] Belenky G, et al. Patterns of performance degradation and restoration during sleep restriction and subsequent recovery: a sleep dose-response study. J Sleep Res 2003;12(1):1–12.

[3] Swanson LM, et al. Sleep disorders and work performance: findings from the 2008 National Sleep Foundation Sleep in America poll. J Sleep Res 2011;20(3):487–94.

[4] Jung CM, et al. Energy expenditure during sleep, sleep deprivation and sleep following sleep deprivation in adult humans. J Physiol 2011;589(Pt 1):235–44.

[5] Carskadon MA, Dement WC. Normal human sleep: an overview. In: Kryger MH, Roth T, Dement WC, editors. Principles and practice or sleep medicine. Saint Louis, MO: Elsevier Saunders; 2011. p. 16–26.

[6] Tononi G, Cirelli C. Sleep and synaptic homeostasis: a hypothesis. Brain Res Bull 2003;62(2):143–50.

[7] Siegel JM. Clues to the functions of mammalian sleep. Nature 2005;437(7063): 1264–71.

[8] Kaufmann C, et al. Brain activation and hypothalamic functional connectivity during human non-rapid eye movement sleep: an EEG/fMRI study. Brain 2006;129(Pt 3): 655–67.

[9] Morrell MJ, et al. Sleep fragmentation, awake blood pressure, and sleep-disordered breathing in a population-based study. Am J Respir Crit Care Med 2000;162(6):2091–6.

[10] Stepanski E, et al. Experimental sleep fragmentation in normal subjects. Int J Neurosci 1987;33(3–4):207–14.

[11] Wang Y, et al. Chronic sleep fragmentation promotes obesity in young adult mice. Obesity (Silver Spring) 2014;22(3):758–62.

[12] Rolls A, et al. Optogenetic disruption of sleep continuity impairs memory consolidation. Proc Natl Acad Sci USA 2011;108(32):13305–10.

[13] Krueger JM, Tononi G. Local use-dependent sleep; synthesis of the new paradigm. Curr Top Med Chem 2011;11(19):2490–2.

[14] Banks S, Dinges D. Chronic sleep deprivation. In: Kryger MH, Roth T, Dement WC, editors. Principles and practice of sleep medicine. Saint Louis, MO: Elsevier Saunders; 2011. p. 67–75.

[15] Bonnet MH. Acute sleep deprivation. In: Kryger MH, Roth T, Dement WC, editors. Principles and practice of sleep medicine. Saint Louis, MO: Elsevier Saunders; 2011. p. 54–66.

[16] Campbell SS, Tobler I. Animal sleep: a review of sleep duration across phylogeny. Neurosci Biobehav Rev 1984;8(3):269–300.

[17] Rechtschaffen A, Hauri P, Zeitlin M. Auditory awakening thresholds in REM and NREM sleep stages. Percept Mot Skills 1966;22(3):927–42.

[18] Colrain IM. The K-complex: a 7-decade history. Sleep 2005;28(2):255–73.

[19] De Gennaro L, Ferrara M. Sleep spindles: an overview. Sleep Med Rev 2003;7(5): 423–40.

[20] Silber MH, et al. The visual scoring of sleep in adults. J Clin Sleep Med 2007;3(2): 121–31.

[21] Kushida CA, Bergmann BM, Rechtschaffen A. Sleep deprivation in the rat: IV. Paradoxical sleep deprivation. Sleep 1989;12(1):22–30.

[22] Walker MP, Stickgold R. Sleep, memory, and plasticity. Annu Rev Psychol 2006;57:139–66.

[23] Louie K, Wilson MA. Temporally structured replay of awake hippocampal ensemble activity during rapid eye movement sleep. Neuron 2001;29(1):145–56.

[24] Pace-Schott EF. The neurobiology of dreaming. In: Kryger M, Roth T, Dement WC, editors. Principles and practices of sleep medicine. Saint Louis, MO: Elsevier Saunders; 2011. p. 563–75.

[25] Wagner U, et al. Sleep inspires insight. Nature 2004;427(6972):352–5.

[26] McGinty D, Szymusiak R. Neural control of sleep in mammals. In: Kryger MH, Roth T, Dement WC, editors. Principles and practices of sleep medicine. Saint Louis, MO: Elsevier Saunders; 2011. p. 76–91.

[27] Scharf MT, et al. The energy hypothesis of sleep revisited. Prog Neurobiol 2008;86(3):264–80.

[28] MacFadyen UM, Oswald I, Lewis SA. Starvation and human slow-wave sleep. J Appl Physiol 1973;35(3):391–4.

[29] Ingiosi AM, Opp MR, Krueger JM. Sleep and immune function: glial contributions and consequences of aging. Curr Opin Neurobiol 2013;23(5):806–11.

[30] Borbely AA, et al. Sleep deprivation: effect on sleep stages and EEG power density in man. Electroencephalogr Clin Neurophysiol 1981;51(5):483–95.

[31] Kripke DF, et al. Mortality associated with sleep duration and insomnia. Arch Gen Psychiatry 2002;59(2):131–6.

[32] Cappuccio FP, et al. Sleep duration and all-cause mortality: a systematic review and meta-analysis of prospective studies. Sleep 2010;33(5):585–92.

[33] Gallicchio L, Kalesan B. Sleep duration and mortality: a systematic review and meta-analysis. J Sleep Res 2009;18(2):148–58.

[34] Martin SE, et al. The effect of sleep fragmentation on daytime function. Am J Respir Crit Care Med 1996;153(4 Pt 1):1328–32.

[35] Ohayon MM, et al. Meta-analysis of quantitative sleep parameters from childhood to old age in healthy individuals: developing normative sleep values across the human lifespan. Sleep 2004;27(7):1255–73.

[36] Wertz AT, et al. Effects of sleep inertia on cognition. JAMA 2006;295(2):163–4.

[37] Jewett ME, et al. Time course of sleep inertia dissipation in human performance and alertness. J Sleep Res 1999;8(1):1–8.

[38] Ribak J, et al. Diurnal rhythmicity and Air Force flight accidents due to pilot error. Aviat Space Environ Med 1983;54(12 Pt 1):1096–9.

[39] Everson CA, Bergmann BM, Rechtschaffen A. Sleep deprivation in the rat: III. Total sleep deprivation. Sleep 1989;12(1):13–21.

[40] Shaw PJ, et al. Correlates of sleep and waking in *Drosophila melanogaster*. Science 2000;287(5459):1834–7.

[41] Borbely AA. A two process model of sleep regulation. Hum Neurobiol 1982;1(3):195–204.

[42] Daan S, Beersma DG, Borbely AA. Timing of human sleep: recovery process gated by a circadian pacemaker. Am J Physiol 1984;246(2 Pt 2):R161–83.

[43] Hursh SR, Dongen HPAV. Fatigue and performance modeling. In: Kryger MH, Roth T, Dement WC, editors. Principles and practices of sleep medicine. Saint Louis: Elsevier Saunders; 2011. p. 745–52.

[44] Consensus Conference P, et al. Joint consensus statement of the American Academy of Sleep Medicine and Sleep Research Society on the recommended amount of sleep for a healthy adult: methodology and discussion. Sleep 2015;38(8):1161–83.

[45] Mirmiran M, Maas YG, Ariagno RL. Development of fetal and neonatal sleep and circadian rhythms. Sleep Med Rev 2003;7(4):321–34.

[46] Taheri S, et al. Short sleep duration is associated with reduced leptin, elevated ghrelin, and increased body mass index. PLoS Med 2004;1(3):pe62.

[47] Foley D, et al. Sleep disturbances and chronic disease in older adults: results of the 2003 National Sleep Foundation Sleep in America Survey. J Psychosom Res 2004;56(5):497–502.

[48] Bliwise DL. Sleep in normal aging and dementia. Sleep 1993;16(1):40–81.

[49] Ancoli-Israel S, Martin JL. Insomnia and daytime napping in older adults. J Clin Sleep Med 2006;2(3):333–42.

[50] Monk TH. Aging human circadian rhythms: conventional wisdom may not always be right. J Biol Rhythms 2005;20(4):366–74.

[51] Hiestand DM, et al. Prevalence of symptoms and risk of sleep apnea in the US population: results from the national sleep foundation sleep in America 2005 poll. Chest 2006;130(3):780–6.

[52] Barger LK, et al. Extended work shifts and the risk of motor vehicle crashes among interns. N Engl J Med 2005;352(2):125–34.

[53] Kramer M. Sleep loss in resident physicians: the cause of medical errors? Front Neurol 2010;1:128.

[54] Rogers NL, Dorrian J, Dinges DF. Sleep, waking and neurobehavioural performance. Front Biosci 2003;8:ps1056–67.

[55] Boyle LN, et al. Driver performance in the moments surrounding a microsleep. Transp Res Part F Traffic Psychol Behav 2008;11(2):126–36.

[56] Torsvall L, Akerstedt T. Sleepiness on the job: continuously measured EEG changes in train drivers. Electroencephalogr Clin Neurophysiol 1987;66(6):502–11.

[57] Dorrian J, Rodgers NL, Dinges DF. Psychomotor vigilance performance: a neurocognitive assay sensitivity to sleep loss. In: Kushida C, editor. Sleep deprivation: clinical issues, pharmacology and sleep loss effects. New York: Marcel Dekker; 2005. p. 39–70.

[58] Van Dongen HP, et al. The cumulative cost of additional wakefulness: dose-response effects on neurobehavioral functions and sleep physiology from chronic sleep restriction and total sleep deprivation. Sleep 2003;26(2):117–26.

[59] Goel N, et al. Neurocognitive consequences of sleep deprivation. Semin Neurol 2009;29(4):320–39.

[60] Harrison Y, Horne JA. The impact of sleep deprivation on decision making: a review. J Exp Psychol Appl 2000;6(3):236–49.

[61] Killgore WD, Balkin TJ, Wesensten NJ. Impaired decision making following 49 h of sleep deprivation. J Sleep Res 2006;15(1):7–13.

[62] Yoo SS, et al. The human emotional brain without sleep—a prefrontal amygdala disconnect. Curr Biol 2007;17(20):R877–8.

[63] van der Helm E, Gujar N, Walker MP. Sleep deprivation impairs the accurate recognition of human emotions. Sleep 2010;33(3):335–42.

[64] Troxel WM, et al. Marital quality and the marital bed: examining the covariation between relationship quality and sleep. Sleep Med Rev 2007;11(5):389–404.

[65] Johnson EO, Roth T, Breslau N. The association of insomnia with anxiety disorders and depression: exploration of the direction of risk. J Psychiatr Res 2006;40(8):700–8.

[66] Dinges DF, et al. Cumulative sleepiness, mood disturbance, and psychomotor vigilance performance decrements during a week of sleep restricted to 4-5 hours per night. Sleep 1997;20(4):267–77.

[67] Goetzel RZ, et al. Health, absence, disability, and presenteeism cost estimates of certain physical and mental health conditions affecting U.S. employers. J Occup Environ Med 2004;46(4):398–412.

[68] Kucharczyk ER, Morgan K, Hall AP. The occupational impact of sleep quality and insomnia symptoms. Sleep Med Rev 2012;16(6):547–59.

[69] Van Dongen HP, et al. Systematic interindividual differences in neurobehavioral impairment from sleep loss: evidence of trait-like differential vulnerability. Sleep 2004;27(3):423–33.

[70] Bachmann V, et al. Functional ADA polymorphism increases sleep depth and reduces vigilant attention in humans. Cereb Cortex 2011;22(4):962–70.

[71] Viola AU, et al. PER3 polymorphism predicts sleep structure and waking performance. Curr Biol 2007;17(7):613–8.

[72] Goel N, et al. DQB1*0602 predicts interindividual differences in physiologic sleep, sleepiness, and fatigue. Neurology 2010;75(17):1509–19.

[73] Thimgan MS, et al. The perilipin homologue, lipid storage droplet 2, regulates sleep homeostasis and prevents learning impairments following sleep loss. PLoS Biol 2010;8(8).

[74] Qureshi AI, et al. Habitual sleep patterns and risk for stroke and coronary heart disease: a 10-year follow-up from NHANES I. Neurology 1997;48(4):904–11.

[75] Ikehara S, et al. Association of sleep duration with mortality from cardiovascular disease and other causes for Japanese men and women: the JACC study. Sleep 2009;32(3):295–301.

[76] Knutson KL, et al. Association between sleep and blood pressure in midlife: the CARDIA sleep study. Arch Intern Med 2009;169(11):1055–61.

[77] Gottlieb DJ, et al. Association of usual sleep duration with hypertension: the Sleep Heart Health Study. Sleep 2006;29(8):1009–14.

[78] Irwin M, et al. Effects of sleep and sleep deprivation on catecholamine and interleukin-2 levels in humans: clinical implications. J Clin Endocrinol Metab 1999;84(6):1979–85.

[79] Dettoni JL, et al. Cardiovascular effects of partial sleep deprivation in healthy volunteers. J Appl Physiol 2012;113(2):232–6.

[80] King CR, et al. Short sleep duration and incident coronary artery calcification. JAMA 2008;300(24):2859–66.

[81] Becker HF, et al. Effect of nasal continuous positive airway pressure treatment on blood pressure in patients with obstructive sleep apnea. Circulation 2003;107(1):68–73.

[82] Spiegel K, Leproult R, Van Cauter E. Impact of sleep debt on metabolic and endocrine function. Lancet 1999;354(9188):1435–9.

[83] Spiegel K, et al. Brief communication: Sleep curtailment in healthy young men is associated with decreased leptin levels, elevated ghrelin levels, and increased hunger and appetite. Ann Intern Med 2004;141(11):846–50.

[84] Markwald RR, et al. Impact of insufficient sleep on total daily energy expenditure, food intake, and weight gain. Proc Natl Acad Sci USA 2013;110(14):5695–700.

[85] Nedeltcheva AV, et al. Sleep curtailment is accompanied by increased intake of calories from snacks. Am J Clin Nutr 2009;89(1):126–33.

[86] Nedeltcheva AV, et al. Insufficient sleep undermines dietary efforts to reduce adiposity. Ann Intern Med 2010;153(7):435–41.

[87] St-Onge MP, et al. Short sleep duration, glucose dysregulation and hormonal regulation of appetite in men and women. Sleep 2012;35(11):1503–10.

[88] Redwine L, et al. Effects of sleep and sleep deprivation on interleukin-6, growth hormone, cortisol, and melatonin levels in humans. J Clin Endocrinol Metab 2000;85(10):3597–603.

[89] Spiegel K, et al. Sleep loss: a novel risk factor for insulin resistance and Type 2 diabetes. J Appl Physiol 2005;99(5):2008–19.

[90] Broussard JL, et al. Impaired insulin signaling in human adipocytes after experimental sleep restriction: a randomized, crossover study. Ann Intern Med 2012;157(8):549–57.

[91] Imeri L, Opp MR. How (and why) the immune system makes us sleep. Nat Rev Neurosci 2009;10(3):199–210.

[92] Mullington JM, et al. Sleep loss and inflammation. Best Pract Res Clin Endocrinol Metab 2010;24(5):775–84.

[93] Thimgan MS, et al. Cross-translational studies in human and Drosophila identify markers of sleep loss. PLoS One 2013;8(4):pe61016.

[94] Irwin M, et al. Partial night sleep deprivation reduces natural killer and cellular immune responses in humans. FASEB J 1996;10(5):643–53.

[95] Spiegel K, Sheridan JF, Van Cauter E. Effect of sleep deprivation on response to immunization. JAMA 2002;288(12):1471–2.

[96] Prather AA, et al. Behaviorally assessed sleep and susceptibility to the common cold. Sleep 2015;38(9):1353–9.

[97] Krueger JM, et al. ATP and the purine type 2 X7 receptor affect sleep. J Appl Physiol 2010;109(5):1318–27.

[98] Krueger JM. The role of cytokines in sleep regulation. Curr Pharm Des 2008;14(32):3408–16.

[99] Patel SR, et al. A prospective study of sleep duration and pneumonia risk in women. Sleep 2012;35(1):97–101.

[100] Hayashi M, Watanabe M, Hori T. The effects of a 20 min nap in the mid-afternoon on mood, performance and EEG activity. Clin Neurophysiol 1999;110(2):272–9.

[101] Donlea JM, et al. Inducing sleep by remote control facilitates memory consolidation in Drosophila. Science 2011;332(6037):1571–6.

Circadian rhythms and sleep–circadian interactions

3.1 CIRCADIAN RHYTHMS

The functioning of the human body can be compared to an assembly line. To run properly and efficiently, the right people have to be present and working at the proper time. There is no reason to have the line running when no one is working. Even at the level of the individual, if one of the assemblers is absent or not synchronized with the other, then the assembly line is disrupted and the product backs up. To prevent this situation from happening, there is a person that coordinates the scheduling so that all the necessary people are in place at the proper time. Under these conditions, the assembly line works efficiently and the most of the products get made.

3.1.1 What are circadian rhythms?

Our bodies need this same type of coordination to use and allocate energy and resources efficiently. The process that synchronizes our body's functions is our circadian rhythm. The word is derived from the Latin *circa* meaning "around" and *dia* meaning "day." Our bodies carry out certain functions at night, when we are typically asleep and other functions during the day when we are awake and interact with the world to learn, socialize, forage, and procreate. Circadian rhythms complete a full cycle in roughly 24 h and keep our body's processes appropriately timed with the outside world. Without the governance of circadian rhythms, these processes may lose their synchronization and their efficiency. This process is entrained by external cues, in particular by external sunlight, to coordinate internal activities with external activities. If this entrainment does not occur, internal daytime functions might peak in the middle of the night. These rhythms are important for our body's and mind's functioning, and failure of these rhythms have been associated with both mental and physical ailments.

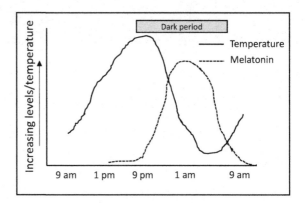

■ **FIGURE 3.1 Examples of circadian rhythms.** The relationship between two circadian rhythms, the core body temperature and the melatonin levels. The temperature rhythm (*solid line*) reaches a minimum point each night during the middle of the sleep period. On the other hand, levels of melatonin (*dotted line*) increase a couple of hours prior to sleep and decline in a way that coincides with the time that one awakens. This pattern repeats itself each 24-h period whether there are external cues or not.

There are some circadian rhythms we are aware of and some we are not [1]. One of the most noticeable rhythms is our sleep and wake cycle. As will be discussed later, the amount of sleep debt is a major factor when we go to sleep, but the circadian rhythms help control the timing of sleep and wakefulness to promote a consistent routine. Another circadian rhythm that one may be aware of is the drop in temperature that occurs each night (Fig. 3.1, solid line). The lowest point of this fluctuation is considered the nadir of this cycle and it occurs regularly at the same time at night of the 24-h oscillation. There are other hormonal rhythms that we do not sense that are executed day in and day out in a 24-h cycle. Our bodies release a hormone, melatonin, with darkness and its release can be hours before one's typical bedtime. It helps to prepare the body for sleep (Fig. 3.1, dashed line). Cortisol, a hormone that promotes activity throughout the body, is also released in a circadian manner toward the end of the night and near the typical rise time of the individual. Since both cortisol and melatonin are hormones, they have the potential to communicate with every cell in the body to coordinate all of the individual circadian rhythms throughout the body. Each of these rhythms continue close to a 24-h cycle in the absence of external environmental cues, but these environmental cues can reset the timing of the 24-h rhythm each day. Thus, they help to set the body's timing for the processes it must carry out.

The circadian rhythms form a clock that controls the internal timing of body functions in a 24-h cycle [2]. The gears of this biological clock consist of a sequential set of biochemical reactions that increase and then decrease the presence of proteins in a cyclic manner. In fact, each cell has its own internal clock that controls cellular functions. All of these individual cellular clocks are coordinated by a master clock in the brain region called the suprachiasmatic nucleus (SCN). The circadian clock does not oscillate in a precise 24-h cycle. Without external cues, these rhythms cycle slightly longer than 24 h and oscillate 4 min longer in males than women [3]. Light is typically the strongest entraining input. Retinal cells detect light and the information is transmitted to the SCN. The molecular clock is flexible, and the environmental cues reset the clock so that the starting point is synchronized with the start of the day by either advancing or delaying the "gears" of the molecular clock to adapt to changes, such as moving across time zones or the change of the seasons. Since we are primarily visual creatures, the circadian rhythms promote wakefulness during the daytime when it best suits our sensory input and sleep at night when there is decreased visual input. Therefore, understanding the molecular mechanism of the clock teaches us something about how to potentially make shift work more tolerable or find the best way to adapt to traveling around the world.

The molecular clock is made up of a series of biochemical reactions that execute sequentially [2]. In this cycle, proteins are generated, interact with one another, are destroyed, and others are downregulated. Each of these individual events takes time which cumulatively adds up to approximately 24 h to execute a complete cycle. A simplified description that should help one understand what a molecular clock looks like is discussed later (Fig. 3.2). Since proteins are responsible for these actions, the actions of individual proteins can be modulated to be faster or slower than the average time taken or they can be manipulated by the environment to help the circadian clock to adapt to new time zones or work schedules.

Step 1, 1 h after the first significant light exposure of the day: Two activator proteins bind to DNA and produce hundreds of proteins that are used during daytime activities. Two of the proteins (we will call them inhibitor proteins) produced actually block the function of the two activator proteins. This is a classic biological negative feedback loop.

Step 2, 6 h after initial light exposure: The activator proteins are active, but the inhibitor proteins accumulate and begin to inhibit the activity of the activators. This increasing inhibition means that the daytime cycle is at its peak. The levels of the proteins that are produced by the activators begin to fall due to the action of the inhibitors. There is a lag between the build up and the effects of inhibition, but the inhibitory process begins here and continues for about next 12 h.

Step 3, 12 h after initial light exposure: The inhibitor proteins continue to block the activator proteins and the levels of proteins for daytime activities fall as the day progresses into the evening and night.

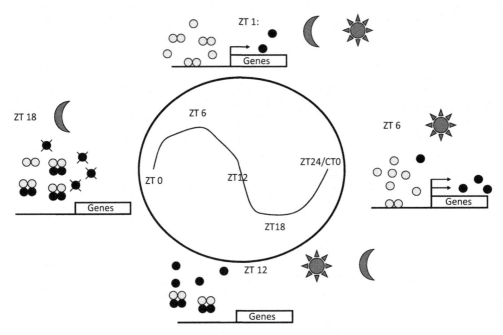

■ FIGURE 3.2 Circadian rhythms at the molecular level. At the beginning of the cycle, ZT 1, the activator proteins (*gray circles*) are able to access the DNA and begin to transcribe thousands of different genes, including the inhibitor protein (*black circles*). By ZT6, the inhibitor proteins are building up because of the activity of activators. At ZT12, the inhibitor protein has inhibited the activity of the activators, which reduces the build up of the inhibitor and other proteins. By ZT18, there is active reduction of the inhibitor protein through degradation of the inhibitor proteins (*X-ed out proteins*), which leads to the increased activity of the activator at ZT1. In the center of the circle is a representation of the oscillation over the 24-h day. The sun and the moon indicate the prevalence of day or night and when both appear at the same time indicates a transition between day and night. The one on the left represents starting point of the transition. See the text for more details.

Step 4, 18 h after light exposure: Levels of inhibitor proteins, that prevent action of activator proteins, decline and the activator proteins now begin to produce proteins that function during daytime activities.

Step 5, Approximately 24 h after light exposure: The levels of proteins cycle back to just about where they were on the previous day at the same time.

3.1.2 Environmental cues that entrain circadian rhythms

The body uses several external timekeepers to establish and set the circadian rhythms. These stimuli include light, meals, exercise, caffeine, and bedtimes and wake times, among others. If these stimuli occur at the same time each day they reinforce the daily cycling of the circadian rhythms.

Out of all these inputs, light exposure is the most potent entrainment stimuli for humans. Light exposure can modulate circadian rhythms to either speed

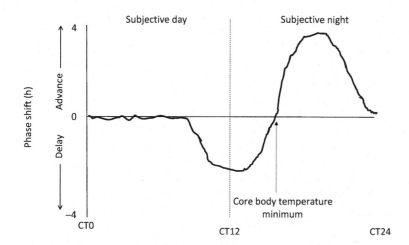

■ FIGURE 3.3 How light can alter the circadian timing or the phase response curve.
Until early in the evening, light does not alter the circadian rhythm. In the early evening, light will delay the circadian rhythms, which results in a later bedtime and rise time. After the core body temperature minimum, the effect of light switches from a delay to an advancing of the rhythms, in which one will waken earlier time than would have been predicted without light stimulation.

up (advance) or slow down (delay) the circadian cycling depending on the time of day that a person is exposed to light, the duration of the exposure, and the intensity of the light [4–7]. If the light exposure occurs before the lowest point of the circadian rhythm (prior to Step 4), then it will induce a delay in the rhythm (Fig. 3.3). One way each of us typically delays their circadian rhythms is by using the modern electric light. Every evening, light exposure throughout the night will suppress the release of the hormone melatonin and delay the circadian rhythms, potentially resulting in a delayed bedtime and rise time. If the exposure to light occurs after the lowest point of the circadian cycle, then the exposure will advance the circadian rhythm and the person will wake up earlier. Thus, light exposure can be used to manipulate circadian rhythms to help align rhythms with the external world.

Given that we are primarily visual creatures, the presence of light is important for our sensory input. In fact, this pathway is so important that there are specific photoreceptors that are dedicated for detecting the presence of light. These photoreceptors send this information to the master circadian regulator in the SCN, which can then communicate the status of external light with the rest of the body. The integration of the intensity and duration of light exposure determines the light's influence on circadian rhythms. In the case of a more intense light exposure, such as daytime sunlight, it takes less time to alter the circadian rhythms. But in dimmer lights, like indoor lighting, it takes a longer exposure time to alter the circadian rhythms, although longer duration of exposure may have more impact than increased intensity [4–7]. Sunlight is one of the best ways to initially set one's circadian rhythms because it is far more intense than indoor lighting. Another method to manipulate circadian entrainment is to use a light box. People can sit in front of a high-intensity light for a short period, often about 30 min to an hour, in the morning to coordinate their circadian rhythms.

It may seem counterintuitive, but many blind people have photoreceptors that are able to communicate the day–night status to the brain even though they don't have image-forming input [1]. Thus, their body stays in sync with the outside world. In some cases, some totally blind people are not able to communicate this information to the brain, which leads to a circadian disruption and may lead to the development of insomnia and other disorders because of a mismatch in their internal body clock with the timing of the external world. These results suggest that light is a key determinant in setting our circadian rhythms. Detecting light through the image-forming portion of the visual system is not the only pathway responsible for setting of the circadian rhythms.

Another stimulus that entrains the circadian system is food consumption [8]. When we consume food, the nutrients are broken down, absorbed into the bloodstream, and shipped throughout the system to the tissues for use. When the brain detects a meal, it marks that time as important and sets the rhythms according to that mealtime. These feeding rhythms are different from those set in response to light. In experiments with animals, researchers shifted the feeding time to the middle of the sleeping period. After a few nights of this consistent schedule, animals began to wake in the middle of the night in anticipation of being fed at that particular time [8]. One recent observation suggests that a consistent and restricted eating schedule is better for one's health even when one consumes the same number of calories [9,10]. This is thought to be because the processes responsible for digestion, energy storage, and energy usage can be coordinated. Restricting the timing of nutrient intake can beneficially consolidate circadian rhythms throughout the body. Therefore, all cells can prepare for an influx of nutrients to store and anticipate times of fasting.

Exercise can be another cue for circadian rhythms. With exercise, the body releases cortisol and norepinephrine, both of which are hormones associated with wakefulness. Routine exercise can also determine the release of these hormones and reinforce the wake-up time for a person. The sleep hygiene recommendation is for one to avoid exercising at night because it will delay sleep onset, though this view may not affect everyone the same way [11,12].

Shifting one's bedtime and caffeine intake can also alter one's circadian rhythm [13,14]. As one shifts their bedtime to a later time, one of the key circadian markers initiates its onset to a later time. This alteration is likely a response to the changing schedule and an adaptation to the external environment that helps coordinate the internal workings with the sleep or wake state that the person will be in at a particular time. It may also be a response to the presence of light later in the evening, which will suppress melatonin secretion. Caffeine also can alter the circadian rhythms and delay them. Whether this occurs through the delay in sleep or through direct effects of caffeine is unclear.

All of these stimuli entrain circadian rhythms to the external world and though they are all integrated and shape the circadian rhythms, light is the strongest input. Therefore, people that are trying to shift their circadian rhythms to fit a new schedule for work or travel should pay particular attention to the timing and duration of light exposure. The next most potent entraining stimulus is meals. Mealtimes can be used to reinforce the circadian rhythms entrained by light.

3.1.3 **Health problems associated with circadian rhythms and shift work**

Since circadian rhythms help prepare the body for what comes next, they coordinate the body's functions. Research has now shown that defects in the circadian system have a large impact on both physical and mental health. Problems with circadian rhythms result in problems in metabolism. Mutations in one of the core circadian proteins leads to obesity, increased levels of cholesterol and serum triglycerides, and diabetes in animal models [15]. These symptoms are part of an emerging health problem called the metabolic syndrome; a set of symptoms that indicates poor metabolic health. In turn, these metabolic alterations can result in obesity, inflammation, and cardiovascular problems, leading to increased risk of heart attack and stroke. People with circadian rhythm problems also exhibit reduced effectiveness of the immune system [16]. Moreover, mental health disorders, such as major depression, schizophrenia, and bipolar disorder, are associated with disrupted circadian rhythms and sleep patterns [17]. Bright light therapy has been effective in treating major depression [18] and anxiety [19]. It remains unclear if the circadian disruption is a side effect of the disease or if circadian disruptions are somehow causative [17]. Different individuals may have different causes, which means light therapy may be very effective in some while ineffective in others.

Shift workers carry both the burden of working against their circadian rhythms and the resulting sleep deprivation. Both of these burdens work against good health [20,21]. Shift workers have a higher likelihood of getting a wide range of health conditions, including gastrointestinal problems, ulcers, diabetes, cancer, hypertension, mood disorders, and cognitive problems [22]. Moreover, shift workers are more likely suffer from insomnia possibly because they are often trying to sleep at a time that their circadian rhythms are promoting wakefulness [23]. Therefore, one explanation for the increased health problems in shift workers is the combination of circadian misalignment and routine sleep deprivation. Evidence suggests that the higher number of years a person works as a shift worker, the higher the risk of many of these conditions [24]. A rotating shift schedule takes the hardest toll on the health and quality of life of the employee. Some individuals adapt well to a stable overnight shift while others do not [25]. Even for those who adapt well, sleep hygiene principles should be employed to help maintain a stable schedule and even then it will take effort to preserve that schedule. With shift work, the person is working while the circadian rhythms would typically dictate that the person should be sleeping. In fact, timed bright light exposure and dark goggles have been used to help people more successfully adapt to a night schedule [25]. Different individuals may be better suited for shift work; however, there may be long-term health effects even in people that adapt to shift work.

3.1.4 Jet lag and daylight savings time

Today's modes of transportation pose a special problem for our bodies, as we do not immediately adapt to the rapid time changes that occur as cars and planes move a person around the continent and the globe. Crossing time zones leads to the condition of jet lag. The essential problem is that the circadian rhythms of the person are timed to the locality from which the person leaves and not to the destination environment. Therefore, the environmental inputs conflict with the person's internal rhythms. When someone travels from the West Coast to the East Coast of the United States, the internal clock for the traveler is set for 3 h earlier than the local environment because they will travel across three time zones. This schedule can make it difficult for the person to wake up for a morning commitment as the internal clock would continue to promote sleep until the typical time that the person would normally wake up, which in this case would be 3 h later. Going to bed can also be a problem because the person may try to go to bed at a reasonable local time, but the person's circadian rhythm may prevent sleep because their rhythm is entrained to the departure locale and is still signaling wakefulness. This situation can result in sleep onset insomnia and the person will not get sufficient sleep. In addition, mealtimes will also continue according to the departure locale's sleep schedule. Traveling in both directions does not have the same impairing effect on the body. Going from east to west is easier than west to east. It is easier to delay meals and bedtime than it is to advance them and get up earlier than the circadian clock would promote.

With time, the circadian clock will adapt to the local time zone. Prior to leaving one can take steps to facilitate this adaptation to the local time of the destination. One helpful step can be to shift meal schedules to those of the destination. In addition, light is the most influential input for the circadian clock. Therefore, if one can begin to rise progressively earlier and be exposed to bright light a few days before the scheduled trip, one can begin to shift the circadian rhythms. Another strategy to advance the circadian rhythms is to progressively wake up 1–2 h earlier each day and be exposed to light. This will enable shifting of the circadian rhythms toward an earlier time (see Section 3.1.2 and Fig. 3.3). The time of day, duration, intensity, and wavelength of the light exposure will determine the impact of light on the circadian rhythms. Moreover, the person's individual makeup will determine the magnitude of these changes. These steps may decrease the amount of time one needs to adapt to their destination.

Circadian rhythms can also be altered using melatonin to advance or delay circadian rhythms. Melatonin and its derivatives are typically taken as a pill and can mimic the natural rise in melatonin. By administering melatonin at specific times of the day, a person may be able to shift their circadian rhythms and adapt to the local environment. The mechanisms and effects of melatonin are discussed in Section 3.2.3. If one wants to advance the clock, melatonin is taken earlier in the evening. In order to delay the clock, melatonin is taken during the morning hours [26]. Adapting one's circadian rhythms can take several days, and it may be important for both worker performance and the amount of quality time one has in the new time zone.

Twice a year much of the world performs a grand jet lag experiment on itself with consequences on health and safety, the biannual custom of pushing the clock forward 1 h in the spring and back 1 h in the fall. Initially conceived of by Benjamin Franklin and implemented just after World War I, the practice of "daylight saving time" (DST) is enshrined in law and essentially puts people through a jetlag experiment without having to go anywhere. This abrupt change results in a mismatch between the environmental light cues and the internal circadian clock. The economic benefits of DST have been questioned, but people do seem to favor longer evenings in the summer. The spring forward typically results in loss of at least an hour's sleep and it is much harder to go to sleep when the circadian rhythms believe it is 1 h earlier either the night before or the night after the time change. The dramatic effects of this shift can be seen in children who have to adjust to the new schedule, in which it is difficult to sleep in the evening and wake up in the morning because of the alteration of the clock time to the external environment. It may take up to 3 weeks for some to readjust their clocks, and some may never habituate to the new schedules [27,28]. During this transition time, incidences of traffic accidents [29], workplace accidents [30], and heart attacks [31] all increase while productivity and cognitive performance decrease [32]. This is likely due to a combination of the change in the circadian alertness and the hour of sleep that is lost in the spring. Thus, DST especially the spring forward, imposes a set of problems that one should be aware of.

3.1.5 Circadian rhythms in the infants and the elderly

Thus far, we have talked about circadian rhythms in healthy mature adults. At the extreme ends of human development, circadian rhythms may not be the same as in healthy people. Infants do not develop the 24-h circadian rhythm for approximately 2–3 months [33]. Therefore, infants will sleep in shorter bouts through both the night and the day. This type of schedule will disrupt the sleep patterns of the parents as the child wakes up intermittently. Parents of newborns should not expect that their newborn would begin to get consolidated sleep through the night before the circadian rhythms are more established. Moreover, there is no set schedule for infants, so it may take a shorter or longer time for one child to develop regular circadian rhythms compared to another.

On the other end of age spectrum, circadian rhythms in the elderly may decline from when they were younger. Cross-sectional studies suggest that the amplitude of circadian rhythms in the elderly is much lower than those of younger adults [1]. This difference in amplitude may have an effect on how coordinated these rhythms are. The lack of consolidated rhythms may result in initiation or exacerbation of health problems. Although this observation has been seen in some studies, it may be that circadian rhythms remain very robust in healthy aged people [1,34]. Thus, it may be that individuals that age well retain robust circadian rhythms or it may be that when circadian rhythms begin to decline that the health of the individual also declines. The reduction of the robustness of circadian rhythms may portend the decline of the individual.

3.2 INTERACTION BETWEEN SLEEPINESS AND CIRCADIAN RHYTHMS

3.2.1 Alertness and vigilance: interaction between circadian rhythms and sleep pressure

In the previous chapter, we presented the importance of sleep debt in determining when someone will go to sleep. As one remains awake, sleep debt continues to increase and will increase the likelihood that the person will fall asleep. For simplicity's sake, we neglected a major contributor to both

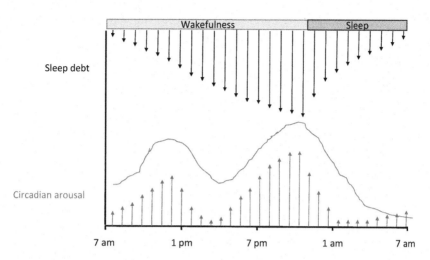

■ **FIGURE 3.4 The relationship between sleep drive, circadian rhythms, and vigilance/sleepiness.** On a typical schedule, sleep debt builds throughout the daytime (*represented by descending black arrows, which put downward pressure on alertness and vigilance*). Longer arrows suggest more sleep debt. Rising arrows estimate the magnitude of the arousal signal over a typical day. Circadian rhythms increase the alertness and vigilance over the day but in a nonlinear manner. The line in the middle represents the relative alertness and vigilance. Thus one can see how these states relate over the day and why performance is not stable over the day and where the danger points might be.

sleepiness and cognitive performance until after the discussion of the circadian rhythms. However, circadian rhythms play a major role by setting the timing of sleep and alertness for an individual.

Circadian activity is associated with wakefulness in humans and circadian output is stronger at certain times of the day compared with others [35,36]. People are often aware of the increased sleepiness and decreased performance levels that occur in the early afternoon. In addition, around bedtime there can be quick transition from being energetic to being sleepy. Both of these are due to the interaction between one's sleep debt and circadian rhythms. The typical pattern of circadian output can be seen in Fig. 3.4 (arrows coming up from the bottom axis). As one awakens, the circadian output builds throughout the morning, increasing the arousal level. As the day proceeds into the early afternoon, the circadian cycle dips and the arousing output is not strong. This is the afternoon lull that one feels and in some cultures this is the daily time for a nap or siesta. As the afternoon progresses, the circadian rhythm output increase sharply and continues to be high in the evening. This is often referred to as a "second wind," and it allows us to function well in the evening. This boost in the circadian output counters the increased sleep debt [37]. At some point in the late evening, the circadian output drops considerably and remains very low throughout the night permitting the mounting sleep debt to induce sleep.

Table 3.1 Typical Combination of Sleep Debt and Circadian Rhythms

Time of Day	Sleep Debt	Circadian Output	Alertness
7 am	Low because the person has just awakened from sleep	Low, but increasing	Relatively high
2 pm	Building from continuous wakefulness	Low because of the typical afternoon decline of circadian output	Low point for daytime activities
8 pm	High from the day's activities	High in the evening	Relatively high
12 am	High from the day's continuous activity	Low to permit sleep	Low and decreasing to the lowest point overnight

Sleep debt and circadian output can be considered as opposing forces, and integrating their influence provides insight into our perceived sleepiness, alertness, vigilance, and performance changes over the day [36]. With a given amount of sleep debt, the performance of the individual will change depending on the time of day because the induced wakefulness changes over the day with the cycling of circadian rhythms (Fig. 3.4). Alertness and vigilance increases over the first part of the day until about just after lunch when circadian rhythms decline. Later in the afternoon, circadian rhythms begin to increase and counteract the build up of sleep debt throughout the evening. At a typical bedtime, the circadian rhythms again decline and remain low throughout the night. Measures of sleepiness, alertness track, and cognitive performance roughly track this integration of sleep debt and circadian rhythms.

Add into this calculation, the impact of sleep debt. Sleep debt will continue to build up as an individual stays awake. Therefore at each time of the day, our vigilance, alertness, and cognitive performance are based on a combination of the amount of sleep debt we have accrued both from that day as well as from prior days of sleep deprivation [38]. While the circadian output counteracts this sleep debt, the arousal signal can be overwhelming by the sleep pressure of a massively accrued sleep debt. The typical combination of sleep debt and circadian rhythms displayed in Fig. 3.4 are summarized in Table 3.1.

The interaction between sleep and circadian rhythm shows a reasonably typical progression over the day can be predicted using mathematical models [35,38]. These models have been set up and validated, and they are currently being used to predict worker performance and fatigue based on work hours, circadian timing, and even changes in time zones for employees, such as pilots. They can help with managing staffing and worker compliance concerns. These models are based on what researchers have learned about how circadian rhythms progress and how sleepiness builds up. These models can be used to determine the appropriate shift rotations and work schedules for a given employee or a given job.

3.2.2 **Shift work**

Shift work poses a special problem in the interaction between sleep and circadian rhythms. Shift work is defined when the performed work is outside of the typical 8 am to 6 pm workday. As businesses run worldwide for 24 h/day, there is a need for people to staff them. Unfortunately, these people are often working against their circadian rhythms. They are trying to stay awake and perform at times when the body is promoting sleep and trying to sleep at times when it is promoting wakefulness. This cycle of trying to work against one's circadian rhythms will often reduce their total sleep time. In total, performance is a combination of time on task, time of day, past sleep history, and the type of task being performed [35]. Shift workers tend to be more prone to poor health, workplace accidents, and poor cognitive performance. These outcomes are a result of both sleep loss and circadian arousal. Unlike jet lag, the local environmental cues do not change and one is working against the natural environmental inputs that reset the rhythm. Some people appear to be able to adapt to this type of schedule while others face a lot of difficulties with the adaptation and may not be well suited for shift work [39]. Some studies have found that older individuals may have a particularly hard time modifying their internal clock to a new shift schedule, though this may not apply to all. Both napping and caffeine can be used as countermeasures. Napping by a shift worker can mitigate the sleep debt that may build up and can dissipate some sleep debt prior to starting one's shift. Caffeine can be used to maintain alertness, though this does not mask all of the effects of sleep debt. One's body also becomes accustomed to caffeine over time and great amounts may be needed for the same result.

3.2.3 **Melatonin**

Melatonin appears to bridge sleep and wake regulation with circadian rhythms. Melatonin is biochemical that is synthesized and released from a portion of the brain called the pineal gland. The rise and fall of melatonin reflects the body's endogenous circadian rhythms but can be modulated to a degree by sleep–wake schedules. The release of melatonin helps coordinate the circadian rhythms that exist throughout the entire body with the master clock in the SCN of the brain [40]. There are two receptors for melatonin that mediate its response in the rest of the body. Melatonin levels begin to elevate around nightfall and remain elevated throughout the night. They then decline back to the baseline in the morning, typically around rise time. Melatonin receptors then transmit melatonin's signal at the cellular level, to modulate sleep and circadian rhythms. Melatonin pills can be used to advance circadian rhythms by taking low doses in the early evening to potentially shift a bedtime to an earlier time point [26]. If melatonin is taken after the lowest point in circadian rhythms, it can push back the time that one wakes up, which is a delay in the circadian rhythms. Therefore, a person's natural melatonin cycle can tell scientists about one's natural circadian rhythm, as well as, externally administered melatonin can alter the natural circadian cycle.

Unfortunately, the levels of melatonin are very sensitive to external light levels, which can suppress its release [4–7]. Even levels of light equivalent to indoor house lights can suppress the increase of melatonin. The photoreceptors responsible for melatonin repression are particularly sensitive to blue light. Many of the devices, such as computers, tablets, televisions, and smartphones, emit enough blue light to suppress melatonin levels. The suppression of melatonin in the evening has the effect of delaying the circadian rhythms and delaying bedtimes. More surreptitiously, tablets and e-readers may have the same problematic effect [41]. In this study, people were asked to read a paper book and the same book from an e-reader. The people that read from the e-reader were up longer, had suppressed melatonin levels, and were made to go to sleep by the investigators, as opposed to the people reading the paper book that went to bed earlier and did not have the melatonin shift. This study demonstrates that technological advances can disrupt sleep in novel ways. In addition, the method by which people carry out what is perceived to be as a relaxing, sleep-inducing activity matters. The delivery method plays a role in sleep regulation. Given the pervasiveness of these devices, it may be wise to rethink their use in the evening and late evening. Individual should be aware of the effects of light so that they can manage their exposure to light to maintain healthy sleep and circadian rhythms.

3.2.4   **Sleep inertia**

There are times that one wakes from sleep still feeling groggy and sluggish. This feeling is often described waking up with cobwebs. In the sleep field, this phenomenon is referred to as "sleep inertia." One hypothesis is that the transitioning from sleep to wakefulness, especially from slow wave sleep, results in decreased cognitive performance [42]. The degree of sleep inertia that one experiences is dependent upon sleep duration, the stage of sleep that one wakes up in, the amount of sleep debt that had built up prior to sleeping, and the time of day that one awakens [42]. Sleep inertia impairs performance on numerous tasks, but may have more of an impact on less-engaging tasks compared to more-engaging tasks [43]. Even after 8 h of sleep, the effects of sleep inertia can be measured up to 2 h after waking [44]. When napping, nap times of 20, 50, and 80 min showed different effects of sleep inertia. The 50 min nap showed the largest effect of sleep inertia likely because the person may wake up from a deeper stage of sleep, such as, slow wave sleep. On the other hand, the 20 min nap showed the shortest effect of sleep inertia because they never entered these deeper stages. Therefore, if one does not go into slow wave sleep or awaken from these stages of sleep, one can reduce the effects of sleep inertia. Additionally, the more sleep deprived someone is, the more pronounced sleep inertia can be. Importantly, the time of day that one takes and awakens from a nap also determines how much sleep inertia one suffers from. Waking up in the middle of the night leads to more sleep inertia than waking in the middle of the day does despite having the same amount of sleep loss [42].

Estimates of how long sleep inertia lasts vary from about 30 min to 2 h depending on task and type of analysis. Different studies have identified that the speed of processing rather than accuracy tend to be more susceptible to sleep inertia, but with sleep deprivation added in, errors in accuracy as well as in processing speed occur [45]. Differences may be due to the type of task, but these results indicate that cognitive processing is affected by sleep inertia. Understanding that sleep inertia can last for up to an hour or more is important for workers who may be pulled into work because of an emergency or illness in a colleague. Therefore, both the employee and the managers should understand the impact that sleep inertia has on performance and how it might be countered.

Circadian rhythms are meant to synchronize the function of the body with the external environment. This helps the body allocate energy efficiently for both, the use of energy in learning and other wakefulness activities, as well as, restoration during times of sleep. The interaction between sleep and circadian rhythms will dictate the vigilance and alertness during the wakefulness period. The response of the individual is based on the prior sleep history, time of day, and current sleep debt. There are detrimental effects when schedules force an individual to work against their circadian rhythms. This misalignment can lead to adverse health and performance effects, such as traffic and work-related accidents. Understanding how to work with one's circadian rhythm when one travels or is engaged in shift work may help in adapting to these situations easier. Powerful entrainers can alter and manipulate the circadian rhythms so that one might maintain their healthiest form and perform at their best potential.

REFERENCES

[1] Czeisler CA, Buxton OM. The human circadian timing system and sleep-wake regulation. In: Kryger MH, Roth T, Dement WC, editors. The principles and practice of sleep medicine. Saint Louis: Elsevier Saunders; 2010. p. 402–19.

[2] Partch CL, Green CB, Takahashi JS. Molecular architecture of the mammalian circadian clock. Trends Cell Biol 2014;24(2):90–9.

[3] Duffy JF, et al. Sex difference in the near-24-hour intrinsic period of the human circadian timing system. Proc Natl Acad Sci USA 2011;108(Suppl. 3):15602–8.

[4] Duffy JF, Czeisler CA. Effect of light on human circadian physiology. Sleep Med Clin 2009;4(2):165–77.

[5] Lewy AJ, et al. Light suppresses melatonin secretion in humans. Science 1980;210(4475):1267–9.

[6] Dewan K, et al. Light-induced changes of the circadian clock of humans: increasing duration is more effective than increasing light intensity. Sleep 2011;34(5):593–9.

[7] Zeitzer JM, et al. Sensitivity of the human circadian pacemaker to nocturnal light: melatonin phase resetting and suppression. J Physiol 2000;526(Pt 3):695–702.

[8] Mistlberger RE. Circadian food-anticipatory activity: formal models and physiological mechanisms. Neurosci Biobehav Rev 1994;18(2):171–95.

[9] Gill S, et al. Time-restricted feeding attenuates age-related cardiac decline in *Drosophila*. Science 2015;347(6227):1265–9.

[10] Hatori M, et al. Time-restricted feeding without reducing caloric intake prevents metabolic diseases in mice fed a high-fat diet. Cell Metab 2012;15(6):848–60.

[11] Buman MP, et al. Does nighttime exercise really disturb sleep? Results from the 2013 National Sleep Foundation Sleep in America Poll. Sleep Med 2014;15(7):755–61.

[12] Irish LA, et al. The role of sleep hygiene in promoting public health: a review of empirical evidence. Sleep Med Rev 2015;22:23–36.

[13] Burke TM, et al. Effects of caffeine on the human circadian clock in vivo and in vitro. Sci Transl Med 2015;7(305):305ra146.

[14] Markwald RR, et al. Impact of insufficient sleep on total daily energy expenditure, food intake, and weight gain. Proc Natl Acad Sci USA 2013;110(14):5695–700.

[15] Turek FW, et al. Obesity and metabolic syndrome in circadian Clock mutant mice. Science 2005;308(5724):1043–5.

[16] Lange T, Dimitrov S, Born J. Effects of sleep and circadian rhythm on the human immune system. Ann NY Acad Sci 2010;1193:48–59.

[17] Jones SG, Benca RM. Circadian disruption in psychiatric disorders. Sleep Med Clin 2015;10(4):481–93.

[18] Lam RW, et al. Efficacy of bright light treatment, fluoxetine, and the combination in patients with nonseasonal major depressive disorder: a randomized clinical trial. JAMA Psychiatry 2016;73(1):56–63.

[19] Baxendale S, O'Sullivan J, Heaney D. Bright light therapy for symptoms of anxiety and depression in focal epilepsy: randomised controlled trial. Br J Psychiatry 2013;202(5):352–6.

[20] Knutsson A. Health disorders of shift workers. Occup Med (Lond) 2003;53(2):103–8.

[21] Matheson A, O'Brien L, Reid JA. The impact of shiftwork on health: a literature review. J Clin Nurs 2014;23(23–24):3309–20.

[22] Vogel M, et al. The effects of shift work on physical and mental health. J Neural Transm (Vienna) 2012;119(10):1121–32.

[23] Drake CL, et al. Shift work sleep disorder: prevalence and consequences beyond that of symptomatic day workers. Sleep 2004;27(8):1453–62.

[24] Schernhammer ES, et al. Rotating night shifts and risk of breast cancer in women participating in the nurses' health study. J Natl Cancer Inst 2001;93(20):1563–8.

[25] Eastman CI, et al. Dark goggles and bright light improve circadian rhythm adaptation to night-shift work. Sleep 1994;17(6):535–43.

[26] Arendt J, Skene DJ. Melatonin as a chronobiotic. Sleep Med Rev 2005;9(1):25–39.

[27] Kantermann T, et al. The human circadian clock's seasonal adjustment is disrupted by daylight saving time. Curr Biol 2007;17(22):1996–2000.

[28] Schneider AM, Randler C. Daytime sleepiness during transition into daylight saving time in adolescents: Are owls higher at risk? Sleep Med 2009;10(9):1047–50.

[29] Hicks RA, Lindseth K, Hawkins J. Daylight saving-time changes increase traffic accidents. Percept Mot Skills 1983;56(1):64–6.

[30] Barnes CM, Wagner DT. Changing to daylight saving time cuts into sleep and increases workplace injuries. J Appl Psychol 2009;94(5):1305–17.

[31] Janszky I, Ljung R. Shifts to and from daylight saving time and incidence of myocardial infarction. N Engl J Med 2008;359(18):1966–8.

[32] Burgess HJ, et al. Can small shifts in circadian phase affect performance? Appl Ergon 2013;44(1):109–11.

[33] McGraw K, et al. The development of circadian rhythms in a human infant. Sleep 1999;22(3):303–10.

[34] Monk TH. Aging human circadian rhythms: conventional wisdom may not always be right. J Biol Rhythms 2005;20(4):366–74.

[35] Balkin TJ. Performance deficits during sleep loss: effects of time awake, time of day, and time on task. In: Kryger MH, Roth T, Dement WC, editors. Principles and practice of sleep medicine. Saint Louis: Elsevier Saunders; 2010. p. 738–44.

[36] Goel N, et al. Circadian rhythms, sleep deprivation, and human performance. Prog Mol Biol Transl Sci 2013;119:155–90.

[37] Edgar DM, Dement WC, Fuller CA. Effect of SCN lesions on sleep in squirrel monkeys: evidence for opponent processes in sleep-wake regulation. J Neurosci 1993;13(3):1065–79.

[38] Hursh SR, Dongen HPAV. Fatigue and performance modeling. In: Kryger MH, Roth T, Dement WC, editors. Principles and practice of sleep medicine. Saint Louis: Elsevier Saunders; 2010. p. 745–52.

[39] Saksvik IB, et al. Individual differences in tolerance to shift work—a systematic review. Sleep Med Rev 2011;15(4):221–35.

[40] Guardiola-Lemaitre B, Quera-Salva MA. Melatonin and the regulation of sleep and circadian rhythms. In: Kryger MH, Roth T, Dement WC, editors. Principles and practice of sleep medicine. Saint Louis: Elsevier Saunders; 2010. p. 420–30.

[41] Chang AM, et al. Evening use of light-emitting eReaders negatively affects sleep, circadian timing, and next-morning alertness. Proc Natl Acad Sci USA 2015;112(4):1232–7.

[42] Tassi P, Muzet A. Sleep inertia. Sleep Med Rev 2000;4(4):341–53.

[43] Santhi N, et al. Morning sleep inertia in alertness and performance: effect of cognitive domain and white light conditions. PLoS One 2013;8(11):e79688.

[44] Jewett ME, et al. Time course of sleep inertia dissipation in human performance and alertness. J Sleep Res 1999;8(1):1–8.

[45] Milner CE, Cote KA. Benefits of napping in healthy adults: impact of nap length, time of day, age, and experience with napping. J Sleep Res 2009;18(2):272–81.

Chapter 4

Sleep hygiene recommendations

Often, people do not realize the number of environmental factors and behaviors that impair our ability to fall asleep or maintain sleep. There are numerous things that we can do to mitigate these sleep disruptions and fall asleep faster. These recommendations are collectively known as "sleep hygiene" practices and can be thought of as cleaning up both the physical and mental environment at or around bedtime. These recommendations can be used to help an employee on both a regular schedule and, importantly, for a person working a shift work schedule. Not every contingency can be addressed here, but the information provided can help one design accommodation to obtain the best sleep possible. Using this information, both management and employees can identify where their practices conflict with what we know about the biology of falling asleep. The information presented here can be supplemented by a wealth of information on specific problems found from trustworthy sources on the web and elsewhere. This information may help move someone from being frustrated with their poor sleep to being able to obtain more, and better, sleep faster.

Difficulties falling asleep or not being able to get needed sleep can be extremely frustrating and deleterious to one's health. In certain circumstances, it can be difficult to fall asleep or sleep for long enough to feel refreshed despite one's best efforts. Sleep difficulties may be brought on by situations at home, rotating shifts, or working long hours. Regardless of the source, the end result is that a person is sleepy, which may affect job performance and the ability to respond to an emergency situation. In other circumstances, a person may not prioritize sleep and voluntarily sleep-deprive themselves without understanding the impact they are having on their health and daytime performance.

Inadequate sleep can impact an individual's health. Reduced sleep has been associated with increased body mass index (BMI) [1], blood pressure [2,3], arterial blockage [4], glucose metabolism [5], being more prone to depression [6], a general decreased quality of life [7,8], and ultimately may lead to a shorter lifespan [9]. In addition, there are effects on a person's daytime functioning including decreased performance levels and an increase in workplace accidents [10], increased absenteeism [10], and mood changes [11] that may affect relationships with coworkers. People often realize they are sleepy during work hours, though they may be unaware of the source of the disrupted sleep or the recommended simple steps to improve sleep duration and quality. The awareness and application of sleep hygiene principles presented here will help to improve the ability to fall asleep and stay asleep. Moreover, the principles of sleep hygiene should also be considered when designing work schedules (see Chapter 10 , Work Shifts) and the work environment to promote optimal wakefulness during shift hours. If the workplace makes time and areas available for employees to sleep, the ideas behind sleep hygiene should be implemented where possible to promote optimal sleep. Though it is impossible to require a good night's sleep, many people are interested in how to improve their sleep. Therefore, even presenting the information and helping someone become more informed about their personal sleep situation will likely improve the overall performance of the workplace based on improved sleep conditions.

The most difficult step may be identifying people that are having difficulties sleeping or whose job performance is negatively impacted by inadequate sleep. At this point, there is no way to easily, objectively, and reliably determine if a person is sleepy using a blood test or other objective metric. There are several groups developing ways to address the problem of detecting sleepiness, including blink detectors. The ideas are being field-tested, and while each has strengths and weaknesses, none is in large-scale use. A fatigue risk management system can fill this void by educating employees on sleep and sleep hygiene. The first group is employees that do not realize that sleep is an important factor in their job performance. These employees may feel that being sleepy is just a rite of passage and do not realize the negative impact that sleepiness may have on their job performance and health. The second group is people that are interested in improving their sleep but they may just not know how. People that take this step may be self-selected individuals that are willing to modify behavior to obtain more sleep. Thus, knowledgeable suggestions at this point may be particularly influential to facilitate improved sleep. The last group is managers that may recognize potential trouble spots and speak to employees about their situation and offer meaningful suggestions on how to improve their sleep situation. Having employees and management that are willing and able to talk about issues of sleep and fatigue will likely improve the workplace environment. Requiring a fatigue risk management system should address all three groups and emphasize the importance of taking adequate sleep seriously.

If a person experiences concerning or persistent serious sleep problems, first and foremost the person should see their primary care physician or a sleep clinician for evaluation. Difficulty sleeping can be caused by numerous health problems that should be treated by a physician. Some examples include excessive daytime sleepiness (EDS) that can be a result of sleep apnea (see Chapter 5, Sleep Disorders), EDS or nighttime wakefulness brought on by medications to treat other maladies, and different cardiovascular, neurological, or other, often painful, disorders associated with sleep disruptions. In each of these cases, one would need treatment from a clinician to properly address the problem. This does not mean that a person cannot play a positive role in improving their sleep under routine circumstances. With some understanding of the physiology of the body and of what factors can disrupt sleep, one can help themselves or someone else improve their sleep. With improved sleep, a person's quality of life, health, attitude, and job performance may improve. Below is a series of recommendations to improve sleep according to the principles of sleep hygiene. Some factors may seem obvious while others are not as intuitive, but rarely are they all implemented to create the optimal conditions to sleep.

4.1 MAKE SLEEP A PRIORITY

In our culture, there is an attitude that sleep indicates you are lazy. The cultural emphasis on work ethic also fosters the pervasive attitude is that one can sleep later and accomplish what one thinks is necessary or what they would like to do rather than prioritizing sleep. Yet, one of the most important aspects of trying to get adequate sleep is to allot enough time to sleep. There are numerous activities that take the place of sleep. Of those, work is one of the major uses of time in people that get less than the suggested amount of sleep [12]. Moreover in modern society, there are ample nighttime activities that can occupy a person. In a survey of over 21,000 people, they stated that they spent more than 55 min of their prebed routine watching television, which may postpone bedtime [13]. In addition, there are video games, Internet, television, reading, social activities, and work responsibilities that may take the place of sleep [13]. Because of this perpetual push to be more productive or to find time for relaxing activities, there is not enough time allocated to sleep.

One example is the parents of young children [14]. They may finally get the kids to bed and then need to take some time for themselves to decompress, to accomplish chores around the house, or to enjoy leisure activity. These may be time-consuming activities that go fairly late into the night. After finally getting to bed and sleeping for a few hours, the kids may wake up early and begin to rouse the parents. The end result of this cycle is a person may get 6 of the 8 h of sleep that they would like to get. The result is repeated sleep restriction that can take its toll on the individual. Therefore, people who do not or are unable to prioritize their sleep consistently get fragmented and inadequate amounts of sleep with health and performance consequences.

The previous example is what sleep researchers call "sleep restriction" and is the most common form of sleep deprivation in society today. In these situations, a person does not get a full night's sleep but does obtain some sleep each night. This can occur voluntarily or involuntarily and will result in the build-up of sleep debt over time. Sleep restriction can have the same consequences as an entire night of sleep loss. One unique feature of sleep restriction is that people that undergo this type of sleep loss do not always feel like they have lost substantial enough sleep to affect their performance. This was demonstrated in an elegant set of experiments performed at the University of Pennsylvania [15]. For 2 weeks, subject spent either 0, 4, 6, or 8 h per night in bed (TIB). TIB of 8 h served as the control for both sleepiness and cognitive performance as evaluated by a reaction time test. Even repeated nights of 6 h TIB over a 2 week period resulted in a significant increase in perceived sleepiness as well as decreased cognitive performance on reaction time and other cognitive tests administered throughout the study. Interestingly, the feeling of sleepiness plateaued after a few days, whereas cognitive performance on the reaction time test continued to decline each day over the entire 2 week period of sleep restriction. At the end of 2 weeks, the subjects' performance on 6 h of TIB was just as bad as if they had gone through a full night of sleep deprivation. Therefore, losing only 1.5–2 h of sleep per night will substantially impair the cognitive performance of a person. As expected, the groups obtaining only 4 h of sleep per night were even worse and the cognitive deficits accumulated faster.

This research reveals three important points about people's understanding of their levels of sleepiness. First up to 10% of the US population gets less than 6 h of sleep [16] a night. Even routine and persistent decrements in sleep may have serious consequences, even if the person does not feel any worse. Second, the cognitive decrements associated with sleep restriction continue to add up as sleep deprivation increases. A severe bout of sleep restriction or a full night of sleep deprivation on top of accumulated sleep debt from previous nights of sleep restriction will result in severe impairment [15]. Lastly, people are very poor judges of their own sleepiness. Even as their performance continued to decrease, the perception by the sleep-deprived subjects did not continue to get worse. Thus, today's schedules may impact one's performance without an individual being conscious of it. If a person does not start by setting aside enough time to sleep, then they will never achieve a fully rested and optimally healthy performing state.

4.1   GOAL: DEDICATE APPROPRIATE AMOUNT OF TIME TO SLEEP

Impediments
- Work deadlines
- Family responsibilities
- Leisure/relaxation activities
- Social activities

Recommendations
- Purposeful effort to adhere to a consistent bedtime

Result
- Better-controlled and more robust circadian rhythms for control of sleep

---

4.2   **LIGHT**

We are primarily visual creatures and our motivated activities during wakefulness primarily occur while there is ample light and sleep when it is dark. Light has two primary methods by which it can modulate waking in people. First, light can reset the circadian rhythms, which will then initiate a program that induces wakefulness in the person according to the circadian schedule [17]. Second, light will stimulate wakefulness independently of the effects on the circadian system [18]. Increased wakefulness will, by definition, disrupt sleep. Thus, light is one of the strongest stimulators of arousal in humans.

We rely on our vision to gather information about the world around us. However, light has a strong influence on sleep and wake physiology in addition to the image-forming role of vision. Light is one of the strongest signals to entrain our circadian rhythm. Our circadian rhythms, or biological clocks, help our body anticipate and prepare for upcoming events in the 24 h day. Much of the mechanism and importance of the biological clock is described in Chapter 3, Circadian Rhythms and Sleep-Circadian Interactions. In fact, this process is so integral to our physiology that there is a separate set of photoreceptors responsible for transmitting light information to circadian centers to synchronize the internal physiology with the daylight schedule [17]. Light of sufficient duration and intensity strikes the photoreceptors to stimulate wakefulness and activity or alter circadian rhythms. Typically, the modulation of circadian rhythms response coordinates activity at the beginning of the day with sunrise. Light advancement of circadian activity can disrupt sleep in the early morning. With this response in mind, it becomes critical to manage light inputs in the work environment, for managing workers after an overnight shift, and for improving the ability to maintain consolidated sleep. An understanding of how light interacts with human biology will help mitigate the effects of light and permit the internal physiology to get optimal sleep.

For workers coming off an overnight shift, the commute home can help stimulate wakefulness. In the morning, the intensity of the sunlight at this time of day can be as much as 10 times as strong as indoor lighting. This is meaningful to someone leaving work at dawn or the early light, as it would increase the likelihood of wakefulness during the commute home. This is beneficial during the commute home, but the intensity of morning light can also be detrimental as the stimulation and circadian effects may delay sleep onset. Thus manipulations of light can have both a positive and negative effect on behavior throughout the day.

The cells that respond to light and stimulate the circadian system respond most strongly to blue light. While other wavelengths of light are stimulatory and can impact circadian rhythms, light in the blue spectrum have an outsized effect by activating these cells and potentially induce wakefulness or alter circadian rhythms. Unfortunately, many of the distractions that keep us awake at night project blue light, such as computer screens, tablets, and televisions. As demonstrated in the time-use survey, many popular prebedtime activities involve a screen that emits blue light close to eyes of the individual [13]. These items continue to be used right up until bedtime, which may increase the time it takes to fall asleep. Thus when one uses these devices, the blue light is activating the arousal system and disrupting sleep. This may frustrate a person that is having difficulties sleeping and initiate a positive feedback loop in which a person cannot fall asleep because of the light stimulation and then becomes concerned about not being able to sleep, which leads to an inability to fall asleep. For blue light coming from a computer, there are now applications that have been developed that remove the blue light emission from the computer display after a certain hour. A solution that would cover all of the sources of blue light is to use glasses that filter out blue light. These measures may have an impact on the ability to fall asleep.

One of the mechanisms by which light works is by suppressing the levels of the hormone melatonin. Melatonin helps prepare the body for sleep and is released from the brain approximately 2–3 h before the anticipated sleep time [19]. Therefore, this hormone helps to set the timing of sleep. Light, even the intensity of light that we use indoors and from computer screens, suppresses the levels of melatonin. Lower levels of melatonin will then decrease the propensity to fall asleep at the appropriate time and may push back sleep initiation [20]. Currently, one use for a melatonin pill is to help people with jet lag as they fly across multiple time zones, especially in a direction in which one needs to fall asleep earlier [21]. The increased melatonin helps the individual to fall asleep earlier in the circadian cycle than they normally would. Therefore, a decrease in melatonin levels would decrease the likelihood that the individual would be able to fall asleep. For a full description of the effects and regulation of melatonin see Chapter 3, Circadian Rhythms and Sleep-Circadian Interactions.

Even in the middle of the night, light can have an effect on our ability to sleep and the timing of waking. At its core, the biological clock is composed of molecules subject to active modulation. Depending on the time of day that light is detected, the effect can either accelerate or decelerate the biological clock. The transition from lengthening the clock to shortening the clock is centered near the low point in body temperature during the night. For example, if the low point in temperature is at 2 am, bright light prior to 2 am will delay the temperature low point and the arousing effects of circadian rhythms. After the 2 am low point, light will accelerate the rhythms so that the arousing signal occurs earlier in the morning. The light stimulation needs to be of sufficient duration and intensity. High intensity light needs only a few minutes to have this physiological effect, while dimmer light needs longer to have the effect. Thus, light exposure during the night may have variable effects on a person's rhythms and can be a powerful manipulator of the human circadian clock, even under short bouts of exposure. For a full description of the circadian response to light see Chapter 3, Circadian Rhythms and Sleep-Circadian Interactions.

Light can also be directly stimulating to humans independent of its effect on the circadian system [18]. There are connections from the nonvisual photoreceptors into brain areas important for sleep. Moreover, increasingly bright light increased the alertness of the people compared with dimmer light [22]. In practice, there are numerous times that people encounter light at a time that it will or could disrupt sleep. Often, one's window coverings are not dark or thick enough to suppress entering daylight. This light can potentially directly arouse the person and also reset the internal clock to activate the arousal pathways governed by the circadian system. For third shift workers, this conflict between light in the bedroom or on the commute home and sleep debt may result in difficulties initiating or maintaining consolidated sleep. Thus, the management of light may aid a person that has difficulty obtaining enough sleep.

To illustrate the complex relationship that humans have with light, it needs to be stated that it is important to have exposure to adequate light as well. This helps to reset circadian rhythms and set the timing of sleep at the appropriate time. As natural sunlight is typically far more intense than indoor lighting, going outside will potentially have a greater effect on the circadian system. This becomes a problem for elderly individuals or any other housebound person because circadian rhythms become less robust and potentially less coordinated without outdoor lighting. Sleep and circadian disruptions have been associated with a host of conditions, including cancer, metabolic problems, and depression [23]. There is a simple treatment, which has worked for some, where one can purchase a light that can deliver an intense amount of white light for 15–30 min first thing in the morning. This will activate and reset the circadian system to coordinate the body's rhythms. Though not as effective as going out into the sunlight, this therapy will help regulate one's circadian rhythms.

Based on the data above, there are several recommendations for one trying to control light exposure to maximize their sleep. The bedroom, especially for a person that needs to sleep during the day, should be as light-free as possible. Blackout curtains and multiple layers over the window should be used to eliminate as much sunlight as possible. Moreover, exposure to light from "screens," such as computers, tablets, phones, and televisions should be limited, especially near bedtime. In particular, shift workers should monitor their exposure at inopportune times to maintain the ability to initiate and maintain sleep. Light, especially high intensity light, should be avoided prior to trying to fall asleep. During the commute home, a person may wear sunglasses to decrease the intensity of the sunlight in order to dampen the effects on the arousal system. These are a couple of modest steps that can be taken to help avoid the unintended outcomes of exposure to different sources of light and improve an individual's sleep.

4.2 GOAL: REDUCE LIGHT EXPOSURE THROUGHOUT OR AROUND SLEEP PERIODS

Impediments
- Sunlight entering sleeping quarters
- Blue light-emitting electronic devices
- Bright indoor lights, especially nearing bedtime or on the commute home

Recommendations
- Blackout curtains for sleeping quarters
- Eye mask for sleeping
- Do not use blue light-emitting electronics prior to bedtime
- Reduce exposure to bright light at night and before bedtime
- Upon waking, increase light exposure to facilitate waking
- Try blue light-blocking glasses

Result
- Permit the increase in melatonin at appropriate times to promote routine sleep timing
- Attempt to align circadian rhythms with work schedule
- Reduce stimulating effects of light during sleep
- Consolidate signals for waking and sleep at the appropriate circadian time

4.3 **CONSISTENT BEDTIME**

One of the most important aspects of sleep is to have a regular and consistent bedtime. This will both help entrain the endogenous circadian rhythms as well as respond to the current cycling of the rhythms. The human body readily adapts to a consistent time schedule as well as to the cycling of the day–night schedule, and there are numerous factors that set circadian rhythms. The first and most influential factor is light, as discussed above. Another system that can entrain circadian rhythms is the eating schedule. It has been shown that if food is presented exclusively during the middle of the primary sleep period, an animal will begin to wake in anticipation of the presentation of food [24]. This is an example of the internal circadian rhythms at work. In addition to light and food consumption, movement/exercise will also entrain an animal's circadian rhythm.

One of the major roles of the circadian system is to coordinate melatonin production. Melatonin secretion increases 2–3 h before bedtime [25]. Therefore, if a person is going to bed at irregular intervals, the circadian system may not be able to coordinate the body's internal environment to the external stimuli and environment. Melatonin secretion will begin and end at different times and not coordinated with the ideal bedtime of the person [26]. This can mean that the onset of sleepiness may be promoted earlier or later than the desired bedtime.

Individuals working the night shift should be particularly aware of maintaining a consistent bedtime, even on days off or weekend days. Constantly shifting the sleep and wake schedule will cause the rhythms to be uncoordinated with the schedule that the person is on. The rhythms will be constantly trying to "catch up" with the shifting schedule. This may be an unavoidable outcome, but it should be a goal to mitigate the transitions that the circadian system attempts to make. For night shift workers, it is easy to transition to a typical daytime schedule on days off, especially under circumstances with children or other social responsibilities, such as visiting with friends or chores that need to be attended to on others' schedules. These situations can put pressure on the individual to attend functions that will disrupt the constant bedtime routine and undo any progress on shifting circadian rhythms to the new schedule [27]. Over the weekend days operating on a normal daytime schedule, the sleep period may drift back toward the middle of the night. This will increase the likelihood that the person will fall asleep inappropriately or becomes inattentive at unanticipated times, such as driving, due to a circadian mismatch or sleep deprivation. Then, when the shift starts again after the weekend, the individual will be off their sleep schedule and may have difficulties sleeping at the appropriate time.

One consequence of the circadian cycle is that the body begins to prepare itself for sleep. Each person has unique make up of circadian rhythms. These have often been used to categorized people as either owls (prefer to be awake at night) or as larks (prefer to be awake during the morning). This may mean that one has a harder time falling asleep during the early night (owl) or maintaining sleep later in the morning (lark). If one naturally has a harder time falling asleep when they are at home, then they will begin to become sleep deprived and have performance issues. One solution to varying circadian rhythms is to find a work schedule that does not conflict with one's circadian rhythms. By doing this, one can take advantage of the natural rhythms so as to maintain consolidated sleep and wakefulness.

4.3 GOAL: OBTAINING A ROUTINE BEDTIME THROUGHOUT THE ENTIRE WEEK

Impediments
- Shift work
- Irregular bedtime (see Section 4.1)
- Weekend "catch-up" sleep
- Social activities

Recommendations
- Purposeful effort to adhere to a consistent bedtime

Result
- Better-controlled and more robust circadian rhythms for control of sleep

---

4.4 BEDTIME ROUTINE

One explanation for the inability to fall asleep has been hyperactivity in the brain [28]. To help settle some of the brain's activity, it has been recommended that a person establish a bedtime routine to help initiate sleep. The bedtime routine can begin to signal to the body that it is time to go to sleep and have the focus move from more stressful and agitating thoughts to calmer thoughts. The first is to have routine that shows the body that it is time to wind down prior to trying to go to sleep. This technique is used in children in which bath and story times in a particular order and with a specific timing are used to cue the child that bedtime is drawing near. This same idea can be applied to adults.

In practice, one could use numerous activities to establish a routine that begins to relax the person. Such activities include a bathing regimen followed by a relaxing book or other wind down activity or hobby that does not frustrate a person. What each individual finds relaxing can be as diverse as there are activities. The recommendations are to find that relaxing routine, start it at least 30 min before bedtime, and adhere to a particular bedtime for best results. In a study of 1600 Finnish men and women, three of the top four perceived sleep-promoting activities were exercise, reading/listening to music, taking a shower/bath/sauna, and maintaining a regular life [29]. In fact, there is experimental evidence that a warm bath in the evening can decrease the time it takes to fall asleep and may have an effect on subsequent sleep [30]. In other words, a routine that included relaxing activities was thought to be the biggest contributors to sleep promotion.

While ideas of how best to fall asleep are very individual, many of the recommendations of what not to do are generalizable across the population. Though television and the Internet are popular evening activities, they will suppress hormones that promote sleep and activate waking pathways (see above). The Internet is an endless source of information that can take up a lot of time searching and exploring. In addition, the subject matter can lead the person to be upset due to the images presented or opinions expressed. Thus, it is advisable to avoid this type of provocation in the time before bed. Video games also pose a challenge. These games are very involved and challenging and are meant for the players to get lost in them. Therefore, these instruments will distract and restrict the sleep of people as they engage in other activities. In fact, it is recommended that there not be a television in the bedroom for many reasons. The first is that one can become occupied with different television shows. Television networks have become efficient at keeping viewers up and watching television. Also, with the increased prominence of services such as Netflix, it is even easier to get hooked on a show or "binge watch" a show, which may keep people up for many hours at a time. In addition, these activities are not recommended as part of the bedtime routine for the stimulatory effects of light (see Section 4.2).

Another recommendation is to avoid emotional conversations right before bed. These conversations can activate and initiate a stressful response. Moreover, with the 24 h news cycle, there are plenty of inflammatory news stories that can activate a person right before bed. This can lead to problems with increased sleep latency. Also, do not eat a large meal before bed (see later). A big meal can disrupt the gastrointestinal system and lead to discomfort that can delay or disrupt sleep. A final recommendation is not to use the bedroom as a workspace, see Section 4.6 for the reasoning behind this idea.

A person can make the bed a stressful place if they associate it with the frustrations of falling asleep. If one is not able to fall asleep within a reasonable amount of time, then they should not stay in bed. Lying awake in bed can lead to frustration with the inability to fall asleep. This relationship can lead to a positive feedback loop in which anxiety builds around the inability to fall asleep, which delays sleep onset, which leads to further anxiety, which delays sleep onset. Thus to reduce this associative anxiety, if one cannot fall asleep in about 20-30 min then one should go into another room and do something relaxing, such as read a book until more sleep pressure builds up and one is able to fall asleep later that night.

4.4 GOAL: GET THE MIND AND BODY RELAXED TO SIGNAL AND MAINTAIN SLEEP

Impediments
- Overactivation of the mind due to stress or contemplation
- Rising in the middle of the night despite inadequate sleep and unable to go back to sleep

Recommendations
- Start a wind-down routine.
- Taking a bath or soothing shower
- Read a book
- Stop working at a comfortable time prior to bedtime
- Reduce emotional agitation

Results
- Falling asleep at the same time each night
- Not lying awake in bed

4.5 NOISE

Environmental noise can disturb sleep. In fact, noise levels have been used to help define sleep. When sleep researchers defined "deeper" sleep, one test was whether the same level of sound would cause an arousal response in the subject. The intensity of the noise established arousal thresholds and defined "deeper" sleep, since it took increased decibel levels to waken a subject compared to "lighter" stages of sleep [31]. Increased arousal thresholds are still used to define sleep. This technique has also been used to experimentally disrupt deeper stages of sleep without causing the person to wake up or changing the actual sleep time [32]. Therefore, noise can cause sleep fragmentation without the person even being aware of it. These same principles transfer to the sleeping environment. There is no way to set a noise level for a particular environment because whether the stimulus rouses a person will depend on what stage of sleep they are in. In other words, a sound of similar decibels may waken an individual in one instance and not in another. Therefore, auditory input may have a variable effect on an individual's sleep and there cannot be an absolute level of sound to permit because it may or may not wake a person up depending on the stage of sleep.

If one must sleep in a noisy environment, there are techniques to help improve the sleep quality. First, one could use ear plugs at night to keep out the sound. Ear plugs are an effective way to block out the sounds, but they are not always the most comfortable way to sleep. Though, eventually one can get used to having the ear plugs in the ear. If these do not work, another remedy is a white noise machine. The idea is to provide noise that will block out or drown out changes in sounds that might disturb one's sleep. If one cannot hear the disrupting sound because of a gentle, repetitive sound put out by the white noise machine, then one might not be woken up by it. There are specific machines that can be purchased that will plug in and have multiple types of sound options that will create an environment that is optimal for sleep. In addition, with the explosion of smart phones, there are free applications that can be downloaded to turn your phone into a white noise machine. These applications also contain multiple types of sounds, including waves crashing on the beach or the sound of a soft rain storm allowing the person to customize the background noise to the one they find the most relaxing. These devices have been shown to increase people's ability to fall asleep and stay asleep in noisy environments.

4.5   GOAL: REDUCE NOISE THAT DISRUPTS SLEEP

Impediments
- Environmental noise

Recommendations
- Ear plugs
- White noise machine

Results
- Neutralize or disguise noise for better sleep

4.6   TEMPERATURE

Temperature is a major factor in the ability for one to go to sleep [29]. As discussed earlier, a warm bath or other moderately warm treatment, such as an electric blanket, just before bedtime may help in falling asleep. Our body temperature changes with sleep onset. It drops just before we fall asleep, and as we go through the night, our body temperature drops by about 0.5°C. It is hypothesized the drop in body temperature is a signal to the body that it is time to sleep. Thus, an increase in temperature through a bath and then the subsequent accelerated cooling when one gets out of the bath or shower may signal the body that it is time to sleep. This dynamic mimics, and may be more pronounced than, the normal thermoregulatory process.

This understanding of physiology may provide the rationale for why the optimal temperatures for sleep are lower. A room with a lower temperature would allow heat to dissipate more quickly and the decline in body temperature would be facilitated. If the room temperature is too warm, then body temperature would not decline as quickly or may not decline to as low of a temperature. If the temperature of the sleeping environment is too hot or too cold, then sleep becomes disrupted because there is a stress response induced that prevents the body from either hypothermia or hyperthermia. Thus, temperature management is an integral part of the sleep process and temperatures outside of a comfortable range may signal danger and induce wakefulness. To obtain an optimal night's sleep, the temperature needs to be in a comfortable range for that individual. Because each individual has unique properties of heat dissipation depending on metabolism, blood flow and distribution of fat (among other things), the optimal sleep environment may be different for each person. That said there appears to be a generally acceptable range from around 65–75°F. Thus any constructed sleeping quarters should contain appropriate environmental controls and should be used to ensure a comfortable sleeping environment.

4.6 GOAL: MAINTAIN IDEAL TEMPERATURE FOR SLEEPING

Impediments
- Bedroom temperature too hot
- Bedroom temperature too cold

Recommendations
- Install ceiling fans or air conditioners in sleeping quarters
- Install heating units in sleeping quarters

Results
- Adjust temperature to permit comfortable sleeping

4.7 STIMULANTS

Stimulants can be loosely defined as something that will induce wakefulness, but they are more often defined as a psychoactive substance that will keep someone awake. Stimulants can come from numerous different sources, and some we take without knowing that there is a stimulant in the product. What make stimulants particularly troublesome for sleep is that often one does not know how quickly the effects of the stimulants will wear off. In part, this is because these stimulants affect each person differently and they are broken down at different rates. Thus, it can be very easy for a stimulant to continue to be active well into the time that an individual would like to fall asleep. This may delay sleep onset or interfere with consolidated sleep. We have listed some commonly consumed products that contain stimulants as well as some timing factors to be aware of with each. This list may help a person if they, or someone they know, are having a hard time falling asleep for an unexplained reason. This is not a complete list, but there may be similarities between the experiences presented and the experiences of the person undergoing sleep difficulties. For a fuller description of sleep promoting and disrupting compounds see Chapter 15, Compounds that alter sleep and wakefulness.

4.7.1 **Caffeine**

Caffeine is the most highly consumed psychoactive drug. It has a generally stimulatory action on humans, which are well known to increase alertness and delay or disrupt sleep. At the beginning of the day, caffeine can be used to increase alertness and counteract the grogginess that is felt immediately after waking. It typically takes 4–6 h for the effects of caffeine to wane but may last longer. This means the caffeine intake should be limited after lunch. Caffeine is found in many commonly consumed food, including most commonly coffee. Substantive levels of caffeine can also be found in some teas, energy drinks, sodas, and chocolate. If one is having difficulties falling asleep, caffeine should be managed during the middle and end of the day. Being "caffeine conscious" can be a simple way to eliminate a sleep-disruptive compound.

4.7.2 **Decongestants**

Decongestants can contain compounds that can disrupt sleep. The same active ingredient that helps clear the congestion also activates similar receptors throughout the body that may increase blood pressure but also induce sleeplessness. These ingredients include pseudoephedrine, phenylephrine, and oxymetazoline. These compounds all ultimately activate the same types receptor and have similar effects. They may be beneficial throughout the day, but if the effects carry over to sleep onset, then they may disrupt sleep. Decongestants labeled nighttime versions often contain diphenhydramine (the active ingredient in Benedryl) or alcohol. Diphenhydramine may have unpredictable effects on sleep, but it generally maintains sleep while the effects of alcohol will be described later.

4.7.3 **Weight loss pills**

With the increase in obesity, diet pills are becoming a more popular form of treatment when diet and exercise regimes do not provide the desired results. Obesity has been associated with many poor health outcomes and decreasing weight has become a focus of medical treatment. To address this problem, some have turned to weight loss pills to aid in managing their weight. Some of these pills have been shown to reduce weight at a safe pace, though it is still a difficult task. One common target for drugs is as an appetite suppressant, which are commonly derivatives of amphetamines. This ingredient is well known to induce wakefulness and disrupt sleep. Therefore, the particular mechanism of action, the timing of administration and whether the person is experiencing sleep problems will all dictate the role that weight management prescriptions can influence sleep. The side effects of this type of "diet pill" may impact the ability of a person to sleep and should be considered as a source of potential sleep disruption.

4.7.4 **Nicotine**

Given that approximately 67 million Americans currently use nicotine, this widely used stimulant has the potential to disrupt sleep in numerous people each night. Nicotine easily passes into the brain and activates specific receptors. It mimics the action of an endogenous compound that stimulates arousal. Therefore, both routine and sporadic use of nicotine may cause sleep problems. It takes routine smokers longer to fall asleep, they have less total sleep time, and they have less of the restorative slow-wave sleep than their nonsmoking counterparts [33]. This is due to the activation of arousal systems, but it is also due to the lack of nicotine ingestion as one sleeps. This withdrawal disrupts sleep. Moreover, the long-term consequences of smoking, such as lung damage, also results in less-satisfying and more fragmented sleep. Sporadic use, such as social smokers who smoke only when at a bar or party, may induce wakefulness due to the atypical boost of nicotine that will disrupt sleep. Often nicotine is used close to bedtime, and therefore can disrupt or interrupt sleep through the stimulation of the brain. Therefore, the recommendation is for people to refrain from nicotine consumption 2–3 h before bedtime. This recommendation carries over to the new electronic cigarettes as well, which is still a nicotine delivery system. Of course, it is recommended that a person quite smoking, but if that cannot be accomplished, then one should consider sleep patterns when considering nicotine ingestion.

4.7.5 **Alcohol**

Alcohol is another commonly used substance that has a complex relationship with sleep. Often people will use alcohol to aid sleep. In fact, alcohol has been shown to decrease sleep latency in humans and even increase slow-wave sleep during the first half of the night [34]. However, during the second half of the night, sleep becomes more fragmented and less restorative. It is also hypothesized that alcohol suppresses rapid eye movement (REM) sleep. Thus, the perception is that alcohol is helping sleep; however, in fact, it is altering sleep architecture through the night and fragmenting sleep as the night progresses. Therefore, alcohol is perceived to be a net negative compared with people that do not use alcohol. In addition, alcohol inhibits a hormone called antidiuretic hormone. This hormone normally helps retain fluids in the body, but when it is disrupted, a person is far more likely to have to urinate. Combine this with the fact that one has drunk a lot of fluid, it is very likely that one will have to get out of bed more than usual in order to visit the bathroom. Each time that one gets up, sleep becomes fragmented and one may not be able to fall back asleep. Also if one is drinking, many popular mixers, such as colas and energy drinks contain caffeine. Therefore, at the end of the night, one may consume quite a bit of caffeine without realizing it, which will further fragment sleep.

4.7   GOAL: MANAGE PHARMACOLOGICS WITH SLEEP SCHEDULES IN MIND

Impediments
- Medication can have stimulant effect that disrupts sleep
- Common foods, over-the-counter compounds, and medications for other conditions have unrealized effects on sleep and wakefulness

Recommendations
- Education program to make individuals aware of the stimulant effects of compounds

Results
- More consolidated sleep
- Less latency to falling asleep

4.8   SLEEPING ENVIRONMENT

The sleeping environment is a critically important part of the sleeping experience, and there are many recommendations that may prevent sleep disruptions. These recommendations will not guarantee that sleep will come easily, but they have been shown to improve the likelihood of sleep and an increased satisfaction with sleep. The first recommendation for the sleeping environment is to remove the television from the bedroom. The television can be a distraction that can affect sleep; it is a source of noise and light that can disrupt sleep. Moreover, if one falls asleep in front of the television, later waking up to turn the television off or reacting to loud or bright moments such as commercials may wake the person and fragment sleep. Furthermore, the television projects blue light, which is known to alter circadian rhythms (see Section 4.2).

Another suggestion is not to use the bedroom or the bed as a workspace. Work often causes feelings of anxiety due to deadlines, mounting work, and stress that accompanies work. When one is trying to fall asleep, the person begins to associate the bed and bedroom with work and all of the stresses that come with it. Thus even on nights in which there is no stressful work to be completed, there may be an associative carry-over where the bedroom is still connected with work anxiety. Therefore, if possible, carry out work in another part of the living space and leave the bedroom for sleep and more intimate moments.

Children, family members, and pets can be a source of sleep fragmentation. It may be hard to recognize disturbances caused by loved children or animals. Young children and sometimes dependant elders often require attention throughout the night. There are feedings, changing diapers, and generally comforting the kids through nightmares, storms, and other situations in the child's life. The priority is to settle the child down so that they can get back to sleep. It is the duty of a loving parent, but it does disrupt one's sleep. Each time, the parent will need to wake and become conscious, address the problem, and then attempt to fall back asleep. This process can take 30 min or more each time it occurs. Depending on the child, these disruptions can occur several times a night. Moreover, parents may try to take some time at night for themselves or to get some chores or work done after the kids have gone to bed. As children often go to bed earlier than their parents and will get up earlier as they dissipate their sleep debt as well as children tend to have earlier acting circadian rhythms. Thus, later bedtimes for parents are accompanied by early rise times with the children. New parents can be accumulating sleep debt for years as they balance work, social life, and child-rearing. Elderly dependants may need much of the same types of care as young children. It can be very demanding to manage one's own life and care for an elderly relative. With both young and elderly dependants, split the duties with another person if possible so that at least one person can get more consolidated sleep. If that is not possible, then it becomes more important to make sleep a priority whenever the opportunity arises.

Another cause of sleep disruption is pets in the house [35]. Pets bring us great joy and pleasure. Research has shown numerous benefits to household pets [36]. One down side of pets is if their schedule conflicts with that of their owner. Animals may need to be fed, have to be let out to relieve themselves, or need to be taken for a walk. Animals with these needs can be quite persistent and waken an individual during sleep. Often these are unrecognized sources of sleep fragmentation possibly because of the love of our pets. It is difficult to change a pet's schedule if it conflicts with one's own, but there may be creative ways to merge schedules with that of the person with sleep difficulties.

The sleeping surface and pillows are another potential source or sleep disruption. Mattresses can be a source of disrupted sleep. An uncomfortable mattress, too soft or too hard, can either result in difficulties falling asleep or waking up before it is time because of pain that is derived from the mattress. With use, the mattress can change, such as sagging, that results in sleep disruptions from discomfort, or leg or back pain. Investing in a new mattress may pay dividends in the long-term through longer, more consolidated sleep episodes that result in more restorative sleep. There are several mattresses that are designed to help couples or individuals sleep better and one may improve sleep depending on the particular individual and sleep problem. Moreover, allergies can disrupt sleep. Antihistamines and other drugs to block allergic reactions can be taken at night to help blunt the response to the allergens. Depending on the type of allergy medication, some may help induce sleep. If allergies to dust mites or other components of the bedding cause a problem, then there are an increasing number of options. First, dispose of and replace the source of the allergen. There are now numerous allergen-suppressing coverings for pillows and mattresses that prevent the buildup of dust mites. Sleeping on comfortable and appropriate bedding can improve the duration and consolidation of a person's night sleep. Though these changes may be expensive, the investment may be the best improvement for one's overall health and well-being.

4.8 GOAL: IMPROVE THE SLEEPING ENVIRONMENT TO ACCOMMODATE OPTIMAL SLEEPING CONDITIONS

Impediments
- Electronic devices in the bedroom
- Using the bedroom as a workspace
- Pets
- Children
- Mattress and pillow comfort

Recommendations
- Use the bedroom for sleep
- Reduce numbers of people and animals in the bedroom
- Reduce electronics use just prior to bed
- Replace uncomfortable bed or pillows

Results
- A comfortable sleeping quarters that promotes sleep

4.9 **PAIN**

Pain is one of the biggest disruptors of sleep [37], making it virtually impossible to obtain consolidated or a sufficient duration of sleep. Because of the many sources of pain, it is hard to generalize to each individual's situation. Some of these sources might be from chronic pain, such as arthritis pain or a bad back. It these cases, staying in the same position during sleep may result in pain at some point in the night and wake a person up. Another source could be acute pain from a number of different sources, such as too vigorous a workout, sunburn, or a bruise from a minor accident, but any presence of pain can be a major disruptor of sleep.

Because pain can waken an individual so readily, it should be dealt with sufficiently, such that a person can sleep at night. In fact, there may be a reciprocal relationship. If the pain is acute, evidence shows that increased sleep may relieve pain faster [38]. Thus, sleep is doing something beneficial to help resolve the issue that is causing the pain. If the pain continually disrupts sleep in an individual experiencing pain, these disruptions may be prolonging pain by decreasing the ability of the body to heal itself.

Of course, there are many ways to manage pain depending on the source. One should consult one's doctor to determine the best treatment regime. For more minor pain caused from acute sources, over-the-counter pain killers may be an acceptable remedy and permit the needed sleep that will help restore the individual to optimal performance levels as well as help the individual heal so that the pain dissipates. This may include aspirin or ibuprofen. These drugs are very good at diminishing or blocking pain, reducing inflammation, and have not been shown to alter sleep. Unfortunately, there are instances when the pain cannot be managed. Under these conditions, there are many physicians that specialize in treating pain as well as very good primary care physicians that can recommend ways to address pain. Importantly, to get a reliably restful night's sleep, one does need to find a way to minimize sleep disruptions that pain causes.

4.9 GOAL: REDUCE PAIN TO PERMIT SLEEP

Impediments
- Acute pain
- Chronic pain

Recommendations
- See a physician for persistent pain
- Temporarily relieve pain with pharmacologics

Results
- Improve duration and consolidation of sleep by alleviating pain

4.10 **DIET**

Diet and eating patterns may also affect sleep [39,40]. A large meal too close to bedtime may result in indigestion or gastric reflux. These can be uncomfortable situations in which the meal consumed can result in discomfort. This discomfort may increase the amount of time that it takes to go to sleep, disrupt sleep during the night or even result in early wake up times because of heartburn. In addition, spicy meals may also alter one's sleep. Especially as one ages, intolerance of certain foods and gastrointestinal response to meals or meal times may increase. These responses are individualistic, and it may take work to determine exactly what factors are causing the discomfort. The time invested to understand one's relationship to the food may improve sleep and quality of life. In general, it is recommended that one eats full meals earlier in the night to provide time for the food to digest before lying down to go to sleep.

A light snack or glass of milk prior to bedtime may benefit sleep [41]. Food intake provides the body with nourishment and energy. Sleep is an active process that requires energy. In fact, there is mounting evidence that if the body feels starved, that sleep will change [42–46]. This is most likely an evolutionarily conserved response to stimulate the animal to go and forage for food as opposed to spending time asleep. Not only is this response found in both humans and other animals, but there is a subset of people that suffer from a condition called night eating syndrome, who will waken in the middle of the night and feel the need to consume food before they can return to sleep [47]. Thus, it may be disruptive to sleep to go to bed really hungry. This line of evidence would reinforce the idea that mother had the right idea that that a glass of warm milk before bedtime will help one fall asleep and stay asleep, though this idea has yet to be directly tested.

One constituent of food is thought to increase sleep and wake [41]. Though it is controversial, the levels of the amino acid tryptophan may correlate with increased sleep duration and deeper sleep. The proposed mechanism of action is that the excess tryptophan is converted to melatonin or serotonin, which can alter the body's chemistry to alter sleep. Melatonin is the hormone that is released prior to when one would normally fall asleep. Though its use is controversial, some believe that they can improve their sleep by taking high levels of melatonin. It is possible that consumption of tryptophan is naturally converted to serotonin or melatonin, which may increase sleep. Outside of serotonin and melatonin, there are numerous other routes that nutrients may increase sleep [41]. It is clear that food can induce numerous pathways that can be both inhibitory and beneficial for a person's sleep characteristics.

4.10 GOAL: MANAGE FOOD TO IMPROVE SLEEP

Impediments
- Poor diet

Recommendations
- Tailor diet to promote sleep
- Reduce consumption of foods that cause gastrointestinal problems such as heartburn

Results
- Dietary promotion of sleep
- Reductions of intestinally-derived sleep disruptions

4.11 NAPS

Napping can be a benefit to an individual's performance, but it has to be managed as to not interfere with sleep that next night. Naps are discussed further in Chapter 8, Naps, which explains many of the principles behind strategic napping [48]. As a general suggestion, a short nap for 15–20 min can restore performance, even if the person does not realize that they feel any better. Moreover, the person would be unlikely to enter slow-wave sleep or REM sleep. Entering these stages of sleep will cause a bout of sleep inertia that makes the person feel very groggy and may result in performance decreases below that of sleep deprived people [49]. In addition, a short nap at an appropriate time of the afternoon will dissipate sleep debt without dissipating too much sleep debt so that one cannot fall asleep at the appropriate time at night. Thus, strategic napping is a way to improve daytime performance in the face of nighttime sleep disruptions. When taking a nap, the recommendations are to limit it to around 15–20 min, try not to take a nap too late in the day or at the low point in the afternoon circadian rhythm to avoid taking too long a nap.

4.11 GOAL: JUDICIOUSLY USE NAPS TO IMPROVE PERFORMANCE WITHOUT INTERFERING WITH NIGHTTIME SLEEP

Impediments
- Responsibilities that make napping difficult
- Sleeping too much during naps

Recommendations
- Limit naps to around 20 min
- Do not nap too close to primary sleep period

Results
- Ability to take advantage of naps while maintaining adequate consolidated sleep

4.12 **BODY POSTURE**

We have a relatively stereotypical position for sleep. In other words, we typically lie down in a comfortable bed to facilitate sleep. On the other hand, one's body posture can be used to help keep one awake. Sitting upright and getting up to move around the room will help to arouse the person and promote wakefulness. With enough sleep deprivation, even these measures may not be enough to keep someone awake, even if they are actively trying to stay awake. It is common to see someone's head bobbing up and down with sleepiness in a meeting while they are sitting upright. Thus, an upright body posture can help to promote wakefulness while lying in the prone position can help increase the likelihood for sleep.

4.13 **EXERCISE**

Exercise is known to have beneficial effects for one's health. The benefits of exercise can carry over to sleep and reinforce one's health through healthy sleep as well. First, exercise is part of a healthy living campaign to reduce one's BMI. BMI attempts to determine lean to fat mass using their weight and height. It is not a perfect metric because there are individuals that do not fit the norms due to body type or workout regimes. But BMI does serve as a relative measure of weight over time in the same individual and can be a good guideline for health and can be correlated as a risk factor for disease in the population. BMI is positively correlated with the onset of sleep apnea as well as other health metrics. Sleep apnea will further fragment one's sleep, reduce the restorative capacity of the sleep, and promote more weight gain. This forms a positive feedback loop in which weight gain increases the likelihood of sleep apnea, which increases one's BMI, which increases the likelihood or severity of sleep apnea. Thus, reducing weight is an important outcome of exercise and has a benefit on the quality of sleep one can obtain.

Exercise also has a beneficial effect on sleep architecture, as those that exercise vigorously have been shown to have more slow-wave sleep periods than people that do not routinely exercise vigorously [50]. This effect may be independent of the effects on metabolism. Slow-wave sleep is thought to be a particularly restorative portion of sleep, and increases in it are thought to be beneficial. One cautionary note about beginning a new exercise program is to not overdo it. One can easily induce a dull pain in muscles that are not used to being used this vigorously. This pain can disrupt sleep, and a good activity can quickly be converted to a detrimental activity.

The timing of the exercise may be a concern for people who are interested in obtaining the best sleep possible. It is thought that vigorous exercise just prior to sleep is detrimental to falling asleep. Exercise is thought to activate the systems of the body and, therefore, vigorous exercise would stimulate the person and make it difficult to fall asleep. Recent evidence has challenged these assumptions regarding late evening exercise [50,51]. The thought is that temperature increase with exercise and the subsequent decrease afterward may be beneficial to sleep. Common advice is to exercise at least a couple of hours before attempting to fall asleep, such as during the morning. Morning workouts would have three potential benefits because one would be engaging in a healthy behavior, reinforcing the onset of the circadian cycle at that time of the morning, and not be potentially increasing the time it takes to fall asleep. The most important piece of advice is to exercise, but if one is having sleep difficulties, then adjusting workout times may improve sleep.

4.13 GOAL: MAINTAIN ROUTINE EXERCISE REGIME WHILE MAINTAINING ADEQUATE SLEEP TIME

Impediments
- Lack of exercise
- Exercising too close to bedtime

Recommendations
- Maintain exercise routine
- If exercise seems to increase latency, try exercising at an earlier time in the day

Results
- Maintenance of weight, health, and sleep

4.14 AGE

As one ages, many of the factors that are discussed above change, potentially in a way that negatively impacts sleep. If one tracks sleep in a population, sleep in the elderly appears to decrease in absolute duration and is more fragmented than at earlier ages. Some have postulated that the reason sleep declines is not due to natural decline in sleep or sleep need, rather sleep declines as a person's health problems increase [52]. As a person ages, the body begins to change and these changes to induce changes in sleep characteristics.

Often it is a combination of many of the factors discussed earlier for the reason that the elderly do not sleep as well as they had when they were younger. Changes that may impact sleep include:

1. Circadian rhythms: It has been reported that the amplitude of circadian rhythms begins to decrease as one ages. Core body temperature does not fluctuate as much in several subsets of the elderly. Also, many

elderly people take daytime naps more frequently than younger individuals, an indicator that circadian rhythms may not be as strong. Many sleep researchers believe that this is not a problem as they calculate the amount of sleep that a person achieves over a 24 h period. Other data suggest that the rhythms in some elderly remain as robust as those in younger individuals [53]. Exposure to bright light helps to entrain the rhythms and home bound people that do not see the bright light of the day may exhibit blunted circadian rhythms. It has also been shown that the elderly may not adjust to abrupt changes in circadian phase as younger individuals.

2. Sleep fragmentation: As one ages, sleep fragmentation increases [54]. If the amplitude of circadian rhythms does decrease, both sleep and wake become less consolidated and there is more sleep during the day and more wakefulness during the night. In addition, bladder control becomes an issue and sleep is fragmented due to an increased need to go to the bathroom during the night. Depending on the state of the sleep debt at the time of arousal, the individual may not be able to go back to sleep after awakening. Therefore, the natural biological process of going to the bathroom will result in a prematurely truncated sleep duration and potentially sleepiness during the following day.

3. Health conditions/medications: With age, there are typically accumulating health problems. The mounting maladies can be part of the sleep disruption. Numerous diseases have been associated with sleep disruptions and poor sleep has been associated with increasing numbers of health problems. Sleep disruptions can result from physiological changes due to disease or from pain induced by diseases that occur with aging. In addition, the medications that one takes to alleviate the illnesses may also disrupt sleep or induce daytime sleepiness.

For some people, adequate sleep can be difficult to obtain. Sleeplessness can be frustrating and may make it even harder to fall asleep or maintain sleep. Increased wakefulness results in people needing to manage their daytime alertness and ability to fall asleep the next night. The tools used to cope with sleeplessness may be unwittingly disrupting their sleep or making achieving sleep more difficult. We have attempted to present some common situations in which people's sleep may be inadvertently interfered with, some of the science behind why these situations might occur, and what one might be able to do to combat the sleep disruptions that occur. It must be stated that each individual has a unique complement of genetic and environmental influences that dictate sleep. Each person's sleep problems may be different with a specific remedy. Each individual must evaluate their own situation and determine where the environmental or internal factors might be impacting their sleep, though some are not obvious sources of sleep disruption. Each individual must also decide the best path to remedy the situation. In some cases there is an approach to counteract the disrupting factor, and in other instances there may not be a clear solution. An awareness of why sleep is delayed or being disrupted can help the person understand and potentially improve their current circumstance. Therefore, each individual would need to develop their own sleep hygiene plan based on their situation and the priorities that they feel are most important in their lives.

REFERENCES

[1] Taheri S, Lin L, Austin D, Young T, Mignot E. Short sleep duration is associated with reduced leptin, elevated ghrelin, and increased body mass index. PLoS Med 2004;1:e62.

[2] Knutson KL, Van Cauter E, Rathouz PJ, et al. Association between sleep and blood pressure in midlife: the CARDIA sleep study. Arch Intern Med (Chic) 2009;169:1055–61.

[3] Mullington JM, Haack M, Toth M, Serrador JM, Meier-Ewert HK. Cardiovascular, inflammatory, and metabolic consequences of sleep deprivation. Prog Cardiovasc Dis 2009;51:294–302.

[4] Grandner MA, Sands-Lincoln MR, Pak VM, Garland SN. Sleep duration, cardiovascular disease, and proinflammatory biomarkers. Nat Sci Sleep 2013;5:93–107.

[5] Spiegel K, Leproult R, Van Cauter E. Impact of sleep debt on metabolic and endocrine function. Lancet 1999;354:1435–9.

[6] Goldstein AN, Walker MP. The role of sleep in emotional brain function. Annu Rev Clin Psychol 2014;10:679–708.

[7] Andruskiene J, Varoneckas G, Martinkenas A, Grabauskas V. Factors associated with poor sleep and health-related quality of life. Medicina 2008;44:206–40.

[8] Van Houdenhove L, Buyse B, Gabriels L, Van den Bergh O. Treating primary insomnia: clinical effectiveness and predictors of outcomes on sleep, daytime function and health-related quality of life. J Clin Psychol Med Settings 2011;18:312–21.

[9] Kripke DF, Garfinkel L, Wingard DL, Klauber MR, Marler MR. Mortality associated with sleep duration and insomnia. Arch Gen Psychiatry 2002;59:131–6.

[10] Kucharczyk ER, Morgan K, Hall AP. The occupational impact of sleep quality and insomnia symptoms. Sleep Med Rev 2012;16:547–59.

[11] Rosen IM, Gimotty PA, Shea JA, Bellini LM. Evolution of sleep quantity, sleep deprivation, mood disturbances, empathy, and burnout among interns. Acad Med 2006;81:82–5.

[12] Basner M, Fomberstein KM, Razavi FM, et al. American time use survey: sleep time and its relationship to waking activities. Sleep 2007;30:1085–95.

[13] Basner M, Dinges DF. Dubious bargain: trading sleep for Leno and Letterman. Sleep 2009;32:747–52.

[14] Hagen EW, Mirer AG, Palta M, Peppard PE. The sleep-time cost of parenting: sleep duration and sleepiness among employed parents in the Wisconsin Sleep Cohort Study. Am J Epidemiol 2013;177:394–401.

[15] Van Dongen HP, Maislin G, Mullington JM, Dinges DF. The cumulative cost of additional wakefulness: dose-response effects on neurobehavioral functions and sleep physiology from chronic sleep restriction and total sleep deprivation. Sleep 2003;26:117–26.

[16] Bin YS, Marshall NS, Glozier N. Sleeping at the limits: the changing prevalence of short and long sleep durations in 10 countries. Am J Epidemiol 2013;177:826–33.

[17] Czeisler CA, Gooley JJ. Sleep and circadian rhythms in humans. Cold Spring Harb Symp Quant Biol 2007;72:579–97.

[18] Hubbard J, Ruppert E, Gropp CM, Bourgin P. Non-circadian direct effects of light on sleep and alertness: lessons from transgenic mouse models. Sleep Med Rev 2013;17:445–52.

[19] Arendt J. Melatonin and human rhythms. Chronobiol Int 2006;23:21–37.

[20] Chang AM, Aeschbach D, Duffy JF, Czeisler CA. Evening use of light-emitting eReaders negatively affects sleep, circadian timing, and next-morning alertness. Proc Natl Acad Sci USA 2015;112:1232–7.

[21] Weingarten JA, Collop NA. Air travel: effects of sleep deprivation and jet lag. Chest 2013;144:1394–401.

[22] Cajochen C, Zeitzer JM, Czeisler CA, Dijk DJ. Dose-response relationship for light intensity and ocular and electroencephalographic correlates of human alertness. Behav Brain Res 2000;115:75–83.

[23] Knutsson A. Health disorders of shift workers. Occup Med (Lond)V 53 2003;103–8.

[24] Mistlberger RE. Neurobiology of food anticipatory circadian rhythms. Physiol Behav 2011;104:535–45.

[25] Cajochen C, Krauchi K, Wirz-Justice A. Role of melatonin in the regulation of human circadian rhythms and sleep. J Neuroendocrinol 2003;15:432–7.

[26] Markwald RR, Melanson EL, Smith MR, et al. Impact of insufficient sleep on total daily energy expenditure, food intake, and weight gain. Proc Natl Acad Sci USA 2013;110:5695–700.

[27] Monk TH. What can the chronobiologist do to help the shift worker? J Biol Rhythms 2000;15:86–94.

[28] Riemann D, Spiegelhalder K, Feige B, et al. The hyperarousal model of insomnia: a review of the concept and its evidence. Sleep Med Rev 2010;14:19–31.

[29] Urponen H, Vuori I, Hasan J, Partinen M. Self-evaluations of factors promoting and disturbing sleep: an epidemiological survey in Finland. Soc Sci Med 1988;26:443–50.

[30] Raymann RJ, Swaab DF, Van Someren EJ. Skin temperature and sleep-onset latency: changes with age and insomnia. Physiol Behav 2007;90:257–66.

[31] Busby KA, Mercier L, Pivik RT. Ontogenetic variations in auditory arousal threshold during sleep. Psychophysiology 1994;31:182–8.

[32] Tasali E, Leproult R, Ehrmann DA, Van Cauter E. Slow-wave sleep and the risk of type 2 diabetes in humans. Proc Natl Acad Sci USA 2008;105:1044–9.

[33] Jaehne A, Loessl B, Barkai Z, Riemann D, Hornyak M. Effects of nicotine on sleep during consumption, withdrawal and replacement therapy. Sleep Med Rev 2009;13:363–77.

[34] Ebrahim IO, Shapiro CM, Williams AJ, Fenwick PB. Alcohol and sleep I: effects on normal sleep. Alcohol Clin Exp Res 2013;37:539–49.

[35] Smith B, Thompson K, Clarkson L, Dawson D. The prevalence and implications of human-animal co-sleeping in an Australian sample. Anthrozoos 2014;27:543–51.

[36] Barker SB, Wolen AR. The benefits of human-companion animal interaction: a review. J Vet Med Educ 2008;35:487–95.

[37] Roehrs T, Knutson KL, Clauw D, Hillygus DS, Smith MT, Webster L. Sleep and pain. In: Sleep and pain: National Sleep Foundation-Sleep in America Poll. http://sleepfoundation.org/sleep-polls-data/2015-sleep-and-pain, 2015.

[38] Smith MT, Haythornthwaite JA. How do sleep disturbance and chronic pain interrelate? Insights from the longitudinal and cognitive-behavioral clinical trials literature. Sleep Med Rev 2004;8:119–32.

[39] St-Onge MP, Roberts A, Shechter A, Choudhury AR. Fiber and saturated fat are associated with sleep arousals and slow wave sleep. J Clin Sleep Med 2016;12:19–24.

[40] Fass R, Quan SF, O'Connor GT, Ervin A, Iber C. Predictors of heartburn during sleep in a large prospective cohort study. Chest 2005;127:1658–66.

[41] Peuhkuri K, Sihvola N, Korpela R. Diet promotes sleep duration and quality. Nutr Res 2012;32:309–19.

[42] Thimgan MS, Suzuki Y, Seugnet L, Gottschalk L, Shaw PJ. The perilipin homologue, lipid storage droplet 2, regulates sleep homeostasis and prevents learning impairments following sleep loss. PLoS Biol 2010;8.

[43] MacFadyen UM, Oswald I, Lewis SA. Starvation and human slow-wave sleep. J Appl Physiol 1973;35:391–4.

[44] Borbely AA. Sleep in the rat during food deprivation and subsequent restitution of food. Brain Res 1977;124:457–71.

[45] Danguir J, Nicolaidis S. Dependence of sleep on nutrients' availability. Physiol Behav 1979;22:735–40.

[46] Keene AC, Duboue ER, McDonald DM, et al. Clock and cycle limit starvation-induced sleep loss in *Drosophila*. Curr Biol 2010;20:1209–15.

[47] Berner LA, Allison KC. Behavioral management of night eating disorders. Psychol Res Behav Manag 2013;6:1–8.

[48] Ficca G, Axelsson J, Mollicone DJ, Muto V, Vitiello MV. Naps, cognition and performance. Sleep Med Rev 2010;14:249–58.

[49] Wertz AT, Ronda JM, Czeisler CA, Wright KP Jr. Effects of sleep inertia on cognition. JAMA 2006;295:163–4.

[50] Irish LA, Kline CE, Gunn HE, Buysse DJ, Hall MH. The role of sleep hygiene in promoting public health: A review of empirical evidence. Sleep Med Rev 2015;22:23–36.

[51] Buman MP, Phillips BA, Youngstedt SD, Kline CE, Hirshkowitz M. Does nighttime exercise really disturb sleep? Results from the 2013 National Sleep Foundation Sleep in America Poll. Sleep Med 2014;15:755–61.

[52] Foley D, Ancoli-Israel S, Britz P, Walsh J. Sleep disturbances and chronic disease in older adults: results of the 2003 National Sleep Foundation Sleep in America Survey. J Psychosom Res 2004;56:497–502.

[53] Monk TH. Aging human circadian rhythms: conventional wisdom may not always be right. J Biol Rhythms 2005;20:366–74.

[54] Vitiello MV, Larsen LH, Moe KE. Age-related sleep change: gender and estrogen effects on the subjective–objective sleep quality relationships of healthy, noncomplaining older men and women. J Psychosom Res 2004;56:503–10.

Chapter 5

Sleep disorders

Inadequate sleep is detrimental to people. Insufficient sleep may result as individuals choose to put off sleep in favor of work and family responsibilities, social obligations, or entertainment. Another possibility is that a person may suffer from a sleep disorder. Sleep disorders are medical conditions that disrupt the duration or consolidation of sleep through a defect in the biological function that regulates sleep or from a condition that requires arousal to adapt to a given condition. They can disrupt sleep by delaying sleep onset, altering the normal sleep stage progression, fragment sleep throughout the night, or reduce sleep duration. Sleep disorders can arise sporadically or derive as a result of another condition.

Given how complex sleep regulation is, it is not surprising there are so many different types of sleep disorders. Sleep disruptions from sleep disorders have many of the same consequences as sleep deprivation. They often result in increased sleepiness, which can reduce workplace performance and productivity. Nearly all have some sort of increased incidence of a medical problem as well. Consequently, if an individual is allotting enough time in bed to get sufficient sleep but is not satisfied with the outcome, it may be wise to consult a physician to determine if there is an underlying sleep disorder. In many cases, there is a treatment that may help the person obtain more satisfying sleep or at least live a full life by alleviating the symptoms of the sleep disorder.

This chapter is meant to inform people about the numerous types of sleep disorders and to describe the consequences that accompany a sleep disorder. Sleep disorders can affect the employee or loved ones, and thus awareness has general applicability. It is not meant to substitute for a sleep clinician's professional opinion on what the problem is and what the appropriate solution might be. We hope that through awareness of sleep disorders, more people may begin to recognize the signs in their own sleep or a loved one's sleep and go to get an official diagnosis and seek treatment that will improve their sleep, health, and cognitive performance.

Due to the breadth of information this chapter attempts to cover as well as space limitations, we are unable to cite all of the original research that has gone into the conclusions presented. There are hundreds of excellent papers by excellent investigators that have identified how these diseases occur and progress, their impact on people, and the best ways to treat them. We apologize that we are unable to individually acknowledge all of these scientific contributions. In the text, we have cited key manuscripts and reviews so that the reader has a starting point if they would like to know more about these topics. If one is further interested in any of the sleep disorders presented here, we encourage the reader to look at these manuscripts and the references within for the primary science that supports their conclusions.

5.1 SLEEP APNEA

Definition: Sleep apnea is one of the most common sleep disorders [1–13]. Sleep apnea occurs when a person has interruptions/pauses (apneas) or exceptionally shallow breathing during sleep (hypopneas). These interruptions can last from seconds to minutes and result in both fragmented sleep and potentially a decrease in blood oxygen saturation. Sleep apnea is a chronic condition in which this process is repeated every night at onset of the disease. Diagnosis of sleep apnea measures both numerous apneas and/or hypopneas per hour known as the apnea–hypopnea index (AHI) in conjunction with symptoms listed as follows under prevalence of sleep apnea.

There are different severities of sleep apnea:

> 0–4 Incidents per hour = no sleep apnea
> 5–14 Incidents per hour = mild sleep apnea
> 15–29 Incidents per hour = moderate sleep apnea
> 30+ Incidents per hour = severe sleep apnea

Sleep apnea has two subtypes. The most prevalent subtype is obstructive sleep apnea (OSA), in which the airway is physically blocked during sleep. The muscles in the back of the throat relax and the airway is either fully occluded (apnea) or partially occluded (hypopnea). With a blocked airway, blood oxygen saturation can begin to drop well below the baseline level. The person then has a brief awakening, the throat muscles contract that opens the airway, and the person falls back asleep without realizing that they had been roused (Fig. 5.1). This cycle can be repeated 30 or more times per hour throughout the night. Therefore, the person achieves less of the deeper stages of sleep because of the constant arousal to open the airway. Thus, the person

(a)

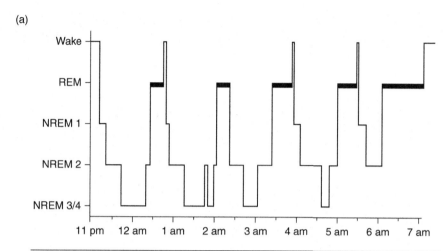

(b)

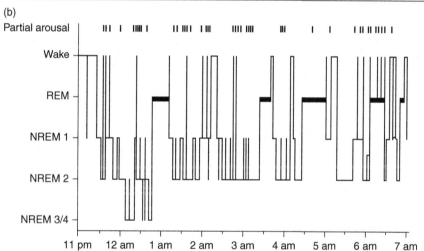

■ **FIGURE 5.1** **Comparison of representative hypnograms from healthy sleepers and individuals with sleep apnea.** (a) Representative hypnogram from healthy sleepers, which shows adequate amounts of deeper sleep (NREM 3/4) in the first half of the night, higher proportion of REM sleep in the second half of the night, and very few wake episodes throughout the night. For a full description of the figure, see Fig. 2.1. (b) Representation of a hypnogram from a person with sleep apnea. Ticks in the "partial arousal" indicate when the EEG was disrupted with an arousal that disrupts sleep brain waves but of which the patient may not be aware. Note that sleep apnea induces frequent wakefulness episodes that fragments sleep.

stays in NREM Stage 1 and 2 sleep primarily and may occasionally achieve Stage 3 sleep. The patient will achieve some rapid eye movement (REM) sleep throughout the night. Despite a full night in bed and what seems like a full night's sleep, a person with sleep apnea may not achieve or feel like they have achieved the full restful and restorative benefits of sleep. They may feel fatigued and tired throughout the day. Over time, a patient with sleep apnea may come to habituate to this type of sleepiness and feel that it is normal. Often with treatment, patients will describe how restful their sleep had been without the interruptions of sleep apnea and how energetic they feel the next morning.

The second subtype of sleep apnea is central sleep apnea (CSA) [2,14]. Under these circumstances, the portion of the brain that initiates breathing will not function properly, which disrupts either the drive or the ability to breathe. As with OSA, CSA results in a reduced breathing, decrease in oxygen saturation, and sleep fragmentation. CSA often results from trauma to the brain that damages respiratory centers or the communication from them. In addition, CSA can result from obesity, certain drugs, muscle or other physiologic problems that reduce breathing, and result from a particular breathing pattern called Cheyne–Stokes respiration, in which breathing waxes and wanes very quickly. The exact prevalence of CSA is unclear. Given the extensive research on OSA, we will use it as the example here. Most of the principles apply to CSA as well.

Prevalence: The prevalence of sleep apnea has been difficult to determine because there is often a long lag between onset and diagnosis, which has averaged up to 8 years. General population reports suggest that OSA is present in 4% of males and 2% of females. Sleep apnea is more prevalent in some populations, including patients with Type 2 diabetics, hypertension resistant to medications, heart failure, and patients that have suffered a stroke. To reduce the time of diagnosis, a bed partner can witness the actual apneas, gasping or choking at night, and describe the snoring should be inclined to suggest seeing a primary care physician. In addition, the patient will often report one of the following symptoms, including excessive sleepiness, nonrefreshing sleep, morning headaches, memory loss, decreased libido, weight gain, increased blood pressure, or irritability. Sleep apnea is found in women, children, and skinny people as well. Thus, all individuals should be on the lookout for sleep apnea in themselves and their cohort of friends and family.

Risk factors: There are numerous risk factors that increase the likelihood of OSA. These include males, increased age, male, African descent, excessive weight [especially a body mass index (BMI > 35)], large neck size, hypertension, narrowed airway, or a family history of sleep apnea. In addition, smoking and alcohol can increase the prevalence of OSA. Sleep apnea is associated with weight gain which may increase the severity of symptoms completing a positive feedback loop. Interestingly, neck size is related to weight, and thus weight may play an outsized role in the presence of sleep apnea. It must be emphasized that women, children, thin and large individuals can all develop sleep apnea, but the risk factors increase the likelihood of sleep apnea.

Consequences: The reason sleep apnea is such a focus is that there are numerous consequences to sleep apnea and likely to be more recognized in the near future. Typically, harsher consequences are seen with more severe AHI and longer duration of disease state. One of the most prominent consequences of sleep apnea is the excessive daytime sleepiness (EDS). EDS impairs cognitive performance as patients report having difficulties staying on task, falling asleep inappropriately at their work space, poor work performance, and memory problems. These outcomes can have adverse effects on career trajectory. Increased AHI leads to increased sleep fragmentation and sleepiness in which there is an increased likelihood that one might fall asleep at inappropriate times and places. Thus, untreated sleep apnea leads to a two to sevenfold increase in car accidents. EDS can also lead to an increase in workplace accidents that can put the employees and others in dangerous situations. Given the duration between onset and diagnosis, patients often feel that they habituate to the sleepiness, but treatment often leads to higher quality objective and subjective sleep.

Moreover, there are serious cognitive consequences with increasing severity of OSA. Patients with OSA show decrements in simple and complex cognitive function, and more difficulties in simple tasks, in which alertness and vigilance are impaired. In addition, higher order thinking, such as planning and foresight and the ability make conclusions or generalizations from a set of facts are worse off in patients with sleep apnea. Moreover, there are some indications that OSA may be associated with depression. These data are preliminary, but there is an increasing link between sleep deprivation and depression. In a small study, the treatment for OSA ameliorated some of the depression present in this population.

The most well-documented health consequences are cardiovascular. Sleep apnea has been associated with heart failure, pulmonary hypertension, transient ischemic attack (TIA), stroke, hypertension, and atrial fibrillation. The cardiovascular symptoms may stem from the increases in total cholesterol, low-density lipoprotein (LDL, the more harmful form of cholesterol), and triglyceride levels or potentially other changes in the body's lipid management. In OSA, biomarkers of cardiovascular problems, such as C-reactive protein and homocysteine, are also elevated in patients. In addition, OSA is associated with an increase in proinflammatory markers, including interleukin-6 (IL-6). All of these changes associated with OSA are detrimental indicators of cardiovascular health.

Another health effect associated with OSA is glucose handling. Patients with Type II diabetes show an increased level of OSA. There is some suggestion that OSA can contribute to developing the adult-onset diabetes, but it is difficult to separate the effects of OSA from obesity. It is unclear whether the OSA leads to obesity and then to insulin insensitivity or directly to insulin insensitivity, but the prevalence of OSA in people with Type II diabetes suggests that OSA contribute to insulin resistance. The cardiovascular, obesity, and glucose handling problems are all part of a new set of systems referred to as the metabolic syndrome. OSA is associated with many of these symptoms of the metabolic syndrome. Ultimately, sleep apnea is associated with a decreased lifespan.

Treatment: The gold standard treatment for sleep apnea is positive airway pressure from either a continuous positive airway pressure (CPAP) machine or a bilevel positive airway pressure (BiPAP) machine. Both machines provide positive pressure to help air pass by the obstruction to permit breathing. In the case of the BiPAP, the positive pressure decreases during expiration, which makes it easier. When using positive airway pressure (PAP) which can be either CPAP or BiPAP, the patient does not suffer the interruptions in breathing, sleep fragmentation, and oxygen desaturation. One of the most replicated results is the reversal of sleepiness that occurs in 1–2 days. Those treated perform better on a driving task. In addition, the patient gets more sleep in the deeper stages of sleep and reports feeling more rested. This reduces the EDS commonly observed and the public safety concerns are reduced with PAP usage.

There is evidence that other health concerns are partly ameliorated through treatment of the apneas and hypopneas in OSA. With CPAP use, there is evidence that there is a modest reduction of hypertension and other cardiovascular metrics, especially after a cardiovascular event such as heart failure or stroke. Cognitive effects of PAP treatment have not shown cognitive improvements with short term use of PAP. In addition, use of PAP reduced depressive symptoms in some people. PAP treatment also reverses some of the glucose metabolism abnormalities. Though it is unclear if PAP treatment will reverse all of the symptoms brought on by OSA, treatment makes a demonstrable difference. Importantly, without treatment there may be a continual decline of health and cognitive parameters.

To be effective, the PAP device must be used properly. The modest effects of PAP treatment may be due, in part, to low compliance rate. There are patients that do not seem to tolerate PAP treatment well. Although industry is working to make devices more comfortable, many choose not to use it. In some cases, positional therapy may be appropriate in which the patient does not sleep in specific positions, such as on their back, in which sleep apnea is promoted. In other cases, an oral appliance may be a better treatment option. This is a device that looks similar to a retainer and functions by pushing on the lower jaw to open up the airway as the person sleeps. This device has been shown to be effective in reducing apnea/hypopnea events but not to the degree of PAP. Another treatment strategy is to have surgery in which part of the throat is removed. This technique may only be suitable for a small subset of people with a particular throat anatomy. It has been shown to be somewhat effective in these cases, but for others, it is not effective. Ideally, patients that suffer from OSA will be able to find the appropriate treatment to reduce the current and future consequences of the disease.

5.2 INSOMNIA

Definition: Insomnia is one of the most widely experienced sleep disorders [15–21]. Many people will experience these symptoms occasionally for a short period of time. In these situations, when the precipitating event has passed, sleep restores to preevent levels. This is called transient or acute insomnia and it is a temporary condition brought on by an external cause. Primary insomnia has a longer duration than just a transient disruption of sleep. In the sleep field, insomnia is defined as sleep that is disrupted in the following ways despite adequate time allocated to sleep opportunity:

1. Difficulty falling asleep
2. Inability to maintain sleep for the full duration
3. Fragmented sleep

In addition, patients with insomnia should have an impairment or sleepiness during the day and it should occur 3 times a week for 1–3 months. These symptoms may result from the activation of the hypothalamic–pituitary–adrenal (HPA) axis, which mediates part of our stress response. Thus, the person is said to be in a hyperaroused state, which makes carrying out sleep difficult. Insomniacs have been found to have higher urinary levels of both cortisol and catecholamines, both of which are released as part of two different arms of the stress response. In addition, aspects of worry about life events may begin to disrupt sleep. This can lead to a person then worrying about not getting enough sleep. The end result is a cycle of rumination that decreases the quality of sleep. Insomnia has been shown to be very persistent with symptoms still present years after the initial onset.

Prevalence: Insomnia is estimated at between 6 and 50% of the population depending on the number of criteria included in the diagnosis. Approximately

30% of the population may have trouble sleeping. When daytime impairment is included the prevalence decreases to 10–15%, and if it is required to be a month in duration, the prevalence drops to 6–10%.

Risk factors: Increased age is a prominent risk factor for increased likelihood of insomnia, possibly due to coexisting medical problems as one ages. Females also have a 40% increased likelihood of insomnia. Female estrogen cycles and menopause constitute an increase likelihood of insomnia symptoms. Poor mental and physical health, psychological distress, anxiety, and depressive symptoms increase the likelihood of the presence of insomnia. In addition, lower socioeconomic status and living alone increase the odds of insomnia. Working nights and rotating shifts also increases the risk of insomnia. Many sleep and circadian rhythm disorders can precipitate or may result in insomnia. Psychiatric disorders, in particular depression, may coexist with insomnia. A familial history of insomnia may predispose a person to insomnia because of either similar genetics or because of sleeping habits instilled by the parents.

Consequences: Insomnia is associated with a substantial decrease in one's quality of life. Insomniacs report lower scores on numerous aspects, including vitality, mental health, general health perceptions, and physical functioning. Insomniacs will complain about daytime functioning, including fatigue or decreased energy, cognitive impairments, especially with attention, concentration, and memory as well as irritability. People suffering from insomnia are two to four times more likely to have an accident in the workplace, decreased productivity, increased absenteeism, and spend 60% more on health care. There are several medical conditions that are found more frequently in patients with insomnia. These include gastrointestinal reflux, pain conditions, and neurodegenerative diseases. In addition, patients with insomnia show hypertension, cardiac events, diabetes, and poorer immune function. Insomnia is a substantial risk factor for the development of a psychiatric disorder. In fact, insomnia may often precede the onset of the mood disorder. Improvements in sleep lead to improvements in the mood disorder. Moreover, insomniacs report more negative interactions with children and partners. Insomnia can be a difficult and persistent affliction, but given the consequences, it is worth the effort to obtain the appropriate amount of sleep.

Treatment: One approach to insomnia has been to combine psychological and behavioral approaches to improve sleep. Insomnia has been hypothesized as either a physiological or psychological activation that disrupts sleep. The initial stressor can lead to an individual becoming stressed about being able to sleep, which will further disrupt the ability to fall asleep or maintain sleep. Therefore, nonpharmacologic approaches have been developed to improve the quality of sleep. These include a combination of methods that will try to give the person a new outlook about their sleep and what bedtime means, provide comforting associations with the bed, relax the person before bedtime, get the person to not stress about sleep, dispense with many of the stressors of the day, build up sleep debt to make sleep easier and integrate sleep hygiene techniques. These techniques have been shown to be effective in improving sleep onset and quality without introducing medications. There are also numerous pharmacologic treatments for insomnia. Depending on the symptoms over the night, how often the symptoms appear, and what other coexisting conditions the person may have, an individual treatment regime that meets the patient's needs should be arrived at in consultation with a physician. These treatments may include drugs such as benzodiazepines which increase inhibition in the brain. There are also more recent drugs, called nonbenzodiazepines that activate a smaller subset of the same class of receptors as the benzodiazepines. These are the commonly prescribed sleep aids, Ambien, Sonata, and Lunesta, and typically have a shorter half-life than the benzodiazepines. The side effects of both classes of medication are sedation and cognitive impairment. Other options include melatonin and the melatonin receptor agonist for sleep onset latency problems, over the counter antihistamines, such as diphenhydramine for sleep maintenance, and other off-label medications that can reduce the symptoms of insomnia. Recently a new approach to sleep problems was developed, which will reduce the wake signal that counteracts sleep signals. Importantly, medications must be tailored to the person's particular sleep problem and what treatment might be most effective and tolerated best given the individual's situation.

5.3 NARCOLEPSY

Definition: Narcolepsy may be the condition most synonymous with sleep disorders [22–27]. It is often lampooned in cartoons and movies. Despite this caricature, it is a serious yet manageable disease. Narcolepsy is defined by the sleep attacks that people have during daytime activities. This can make it dangerous to drive a car and challenging to sit through meetings during the day. Ironically, it can be difficult for people that suffer from narcolepsy to sleep at night. Moreover, narcolepsy is often accompanied by another condition known as cataplexy. With cataplexy, emotional triggers will cause muscle relaxation to the point that the person will collapse or feel as though they can no longer hold themselves up. For example, if friends tell a funny joke or story, a person with narcolepsy may feel weakness in the legs and difficulty standing. The symptoms of narcolepsy are manageable with pharmacological treatments and with care, a person can live a normal lifespan.

There is mounting evidence that the root cause of narcolepsy is failure in communication of the neuropeptide, orexin (also known as hypocretin). This brain communication system helps to maintain stable wakefulness. Therefore, if either the cells that produce this neurotransmitter or the cells that receive that signal do not function properly, then wakefulness will not be consolidated and one may fall asleep much more easily at unwanted times. One explanation for the disease is that there is an autoimmune response that specifically kills these neurons. Once the source of orexin is gone from the brain, sleep and wakefulness states become unstable and the symptoms of narcolepsy present. The mechanism that determines whether a person does or does not get narcolepsy is more complicated than testing for a particular biomarker, but people with a particular genetic marker are more susceptible to the autoimmune response.

Since narcolepsy is such a rare disease with general symptoms and a gradual onset, it can be difficult for a clinician outside the sleep field to recognize narcolepsy. It is reported that it can take up to 7 years from the onset of symptoms to diagnosis. Emerging work may help in this process as one group has identified low levels of orexin in the cerebrospinal fluid (CSF) as a molecular biomarker associated with narcolepsy. A molecular indicator may help in the accurate diagnosis of narcolepsy.

Prevalence/risk factors: Narcolepsy occurs in about 1 in 2000 people (about 0.05% of the population). The symptoms of narcolepsy develop in men and women equally and begin to set in about the late teens or early 20s of females and in the 30s for males, though it can develop at any age. Most of the cases of narcolepsy are with cataplexy, though 10–50% are without cataplexy. There may be a familial component to narcolepsy, as there is a 1–2% risk that a first-degree relative will have narcolepsy as well. Genetic data may support this hypothesis as a particular genetic marker, DQB1*0602, is linked to an increased rate of narcolepsy, though research indicates there are likely to be other genetic loci associated with narcolepsy. Not everyone with the DQB1*0602 marker develops narcolepsy; therefore, it is likely that there is some undiscovered event or trigger that leads to the development of narcolepsy.

Consequences: The most serious consequences of narcolepsy are the increased likelihood of having a sleep attack during an inopportune or even dangerous moment, such as while driving. The sleep attacks can also occur in social, work, or school situations, in which the person can be put in an awkward, uncomfortable, or disadvantaged position that affects their performance. Additionally, cataplexy can add to the discomfort in dealing with others. Without treatment, the unpredictability of behavioral state can be debilitating and result in a lack of confidence, dependence, and social isolation. In addition, people with narcolepsy may see increased weight gain.

Treatment: The symptoms for narcolepsy can be managed using behavioral modifications as well as medications to control the symptoms and try to consolidate the sleep and wakefulness periods and reduce the cataplexic events. One suggestion is to schedule two to three naps during the day to prevent sleep pressure from building up. Therefore, there is less chance for the patient to fall asleep at an unwanted time. In addition, medications can help narcoleptics live a full life with the disease. Sodium oxybate has been FDA approved for the management of cataplexy and it reduces sleep fragmentation in narcoleptic patients by increasing the deeper SWS and reducing Stage 1 sleep and the number of arousals. Medications such as tricyclic antidepressants or norepinephrine reuptake inhibitors can help to reduce the cataplexic events and suppress REM sleep or increase neurotransmitters responsible for waking. Other medications may help consolidate sleep and wakefulness to varying degrees of success. Thus, there are treatment options that can help a person manage narcolepsy.

5.4 RESTLESS LEG SYNDROME (RLS)/ WILLIS–EKBOM DISEASE (WED)

Definition: RLS (or WED) is a condition in which patients describe an uncomfortable or unpleasant sensation in the legs that causes them to voluntarily move their legs or less frequently other muscles, to temporarily alleviate the uncomfortable feelings [28–31]. Patients have a hard time to accurately describe the sensation, but it has been described as feeling like pain, electric shocks, tingling or creepy sensation. These sensations are not uniformly present over the day, as they are more prevalent in the evening or night time with increased immobility and thus tend to peak near bedtime. The sensation and voluntary movement required to pacify the sensation can delay sleep onset or fragment sleep resulting in shorter sleep duration. Though RLS can be thought of as a movement disorder, it becomes a sleep disorder because the condition will delay sleep onset, reduce sleep duration, and ultimately increase sleepiness and decrease quality of life. RLS can occur less than once per year in some patients whereas it can occur every night in others depending on whether one's symptoms are mild, moderate, or severe.

RLS can be divided into primary, which is where RLS is the primary disease, and secondary, in which RLS symptoms appear because of another disease. Conditions that can precipitate RLS include renal disease, iron deficiency, neuropathy, use of antidepressants, or numerous other conditions. In secondary RLS cases, the RLS is usually resolved if the primary disorder is successfully addressed. In primary RLS, it has been hypothesized that there is a failure of dopaminergic neurons or reduced dopamine signaling that results in the symptoms of RLS. This hypothesis postulates that there is a decrease in dopaminergic signaling that reduces the inhibition of motor neurons during periods of relaxation. With the inhibition lowered, there will be more movement stimulated in the limbs. Data support this hypothesis. First, at periods of peak RLS events dopamine levels are at their lowest. Second, dopamine agonists tend to reduce symptoms of RLS. Third, dopamine is reduced in animals that show symptoms of RLS. Fourth, in model organisms, the symptoms of RLS can be induced by eliminating a specific population of dopaminergic neurons.

The other major finding is the association of RLS with low blood iron levels in 31% of people with RLS. For patients, the severity of the RLS negatively correlated with blood iron levels. Genetic susceptibility for RLS has also been linked with lower plasma iron levels. The link between these two findings could be that iron is a cofactor for the enzyme that synthesizes dopamine. There was lower amounts of iron found in the substantia nigra, an area of the brain that produces dopamine. Therefore, when the iron cofactor is low, the dopamine synthetic enzyme may not function efficiently, and dopamine levels will be reduced. Low iron also reduces signaling through a critical dopamine receptor. Evidence collected to date suggests that reduced dopamine signaling in the brain release the inhibition of motor neurons during the early part of the night, which results in more stimulated voluntary movement of the muscles.

Prevalence: The prevalence of RLS in the population is estimated to be between 5 and 8.8%. Of those, 50% have mild or infrequent symptoms. In children, the estimated prevalence is 2%. The prevalence increases up to about age 60 and then prevalence begins to decrease. Symptoms can disappear for some time and then reappear, possibly because of some of the risk factors below. Onset of RLS is seen in both children and as well as adults. The average age of onset for primary RLS is between 33 and 35 years old. Onset for secondary RLS is between 42 and 52 years old. 40% of people with RLS report having symptoms prior to age 10.

Risk factors: There are numerous factors associated with the increased incidence of primary RLS. Incidence and severity of RLS increases with age. Symptoms can be exacerbated by stress, fatigue, and sleep deprivation. In adults, women are twice as likely as men to exhibit symptoms of RLS, but there is an equal distribution between boys and girls. There may be a higher incidence in people of Northern European descent compared with Asian descent. Moreover, 20% of pregnant women will have symptoms of RLS during their pregnancy. Low levels of iron are a risk factor. People with cardiovascular problems or coronary artery disease exhibit twice the incidence of RLS compared to patients without. RLS is also associated with obesity. Numerous drugs can exacerbate symptoms of RLS, including alcohol, caffeine, and tobacco as well as numerous medications, including medication to treat high blood pressure and depression. If one sees the sudden onset, one must talk with a physician to determine if medications for other diseases are aggravating the symptoms of RLS.

The presence of RLS is also associated with genetic factors. RLS appears to run in families as 18.5–92% of people with RLS have family members with symptoms. Human genetic studies have identified numerous genetic alleles associate with RLS. In one study, 6 genetic loci were associated with RLS that could account for about 80% of the risk of RLS. Interestingly, one gene is known to be involved in the body's handling of iron. Thus genetic evidence support the hypothesis that genetics can predispose a person to RLS and that iron handling may be important in RLS physiology.

Consequences: RLS can severely disrupt sleep and substantially reduce the quality of life. These symptoms can vary from mildly annoying to very disruptive. Individuals with moderate to severe RLS sleep on average 5.5 h, have longer sleep latencies, and a higher arousal index, suggesting their sleep is more fragmented and they have more difficulty sleeping. Patients report increased daytime sleepiness, fatigue, anxiety, major depression, hypertension, heart disease, and importantly, impaired daytime functioning. The sleep and health effects decrease the quality of life, and 50% of RLS sufferers report that the sleep disruptions and the health/performance decrements impact their daily activities and personal relationships.

Treatment: There are numerous approaches to treating RLS. Often the first place to start are nonpharmacologic treatments, especially with mild or infrequent symptoms. Behavioral therapies can include changes such as regular exercise, good sleep hygiene, heating pads, warm or cool baths, massage, or mental distraction. Behavioral approaches may be helpful in up to 80% of RLS.

For the 20% of people that behavioral therapy does not alleviate symptoms, there are several pharmacologic options to treat RLS. The first-line pharmacological treatment for RLS is to increase dopamine signaling. Drugs have been designed to boost the low dopaminergic signaling found in RLS patients. Another set of drugs targets specific dopamine receptors to activate dopamine signaling just in these receptors. This approach has the advantage of reduced augmentation, a side effect in which the symptoms actually get worse with increased dose or duration of treatment. If augmentation occurs or the patient's RLS symptoms do not respond to initial treatments, other pharmacologic agents also work. One possibility are agents that increase inhibition throughout the nervous system, such as medications that help reduce seizures. In addition, opiates reduce the symptoms of RLS, but they have a high risk of dependence.

Treatment of an underlying physiological condition may also alleviate symptoms of secondary RLS. In the case of low iron, taking exogenous oral iron may be sufficient. Iron is often taken with vitamin C to improve gut uptake. If a person does not have an iron deficiency, then gut absorption will not permit an excess of iron uptake and iron therapy will not alleviate the symptoms. In the case of other sources of secondary RLS, addressing the underlying problem will often resolve the RLS symptoms.

5.5 SHIFT WORK DISORDER

Definition: This particular sleep disorder results from the 24-h production culture that has arisen during the industrial age [32–35]. Shift work is defined as any work time that is outside of the conventional 9 to 5 workday. As employees work around the clock, they may labor at a time that is opposite of the body's clock, especially during overnight shifts. As discussed in Chapter 3: Circadian Rhythms and Sleep-Circadian Interaction, our bodies have an endogenous rhythm that promotes consolidated wakefulness during the daytime and promotes sleep during the night. For a person that works the overnight shift, the work and circadian schedule may conflict and force an individual to attempt obtain much needed sleep at a time when their body is sending wakefulness signals leading to fragmented, shorter duration, or nonrestorative sleep. These conditions may be repeated over several nights and weeks as the overnight shift schedule continues. Thus, shift work becomes a sleep disorder when one's work schedule begins to interfere with obtaining adequate sleep. People with shift work disorder will complain about insomnia or excessive sleepiness temporally associated with a recurring work schedule that overlaps the usual time for sleep. These symptoms occur over the course of at least 1 month because of a circadian and sleep-time misalignment. These workers often show daytime insomnia and therefore have excessive nighttime sleepiness that can influence their ability to perform throughout their shifts. They often will exhibit a short sleep latency but sleep is truncated before a morning shift because of waking by an alarm. The lost sleep typically reduces the amount of time that the person spends in Stage 2 and REM sleep while SWS amounts are preserved. The outcome is that sleepiness builds up in the individual and performance declines leaving them prone to accidents. Even if one is on a stable night shift, daytime sleep does not appear to improve after several night shifts. This highlights how strong the pull is to match our internal clock with the external day–night schedule. Individuals tolerate the move to shift work differently and identifying the ones that can stand this misalignment will benefit the worker and the organization.

Prevalence: Approximately 20% of the workforce carries out shift work. Of those people, it has been estimated that between 5 and 10% of the total number suffer from shift work disorder.

Risk factors: Being on a shift work schedule is the risk factor for this disorder. Individuals can lose between 1 and 4 h from their typical sleep duration by being on a rotating shift schedule, especially when moved to the night shift. These shifts result in the employee trying to sleep during the daytime, when wakefulness is being promoted and trying to work during the night, when the body is promoting sleep. On a rotating shift, the body never has an opportunity align one's circadian rhythms with the work schedule, if possible. Maintaining a consistent shift has been shown to have minor benefits, though the natural rhythms of the person likely plays a role in how an individual responds to the rotating shifts. Older individuals may have a harder time being placed on shift work, especially if one has never been on a shift schedule before. Gender and whether you are a morning person will have a negative impact. There is a genetic component as well. People with a genetic variant of the *Period 3* genetic locus are more prone to insomnia with sleepiness on a shift schedule. Thus, many factors can increase the likelihood one will get shift work disorder.

Consequences: The consequences of shift work disorder are inadequate sleep due to circadian-work shift misalignment. This can lead to decreases in productivity compared to daytime performance. Importantly, the sleepiness can pose a safety problem as sleepiness results in more workplace accidents. In addition, sleepiness can result in more driving accidents on the commute on the way to and from work. Increased sleepiness, including that from shift work, has been linked to numerous health problems, especially in shift workers. Not only can it exacerbate preexisting depression, bipolar depression, and substance abuse, but it is also associated with diabetes, obesity, hypertension, impaired reproductive health, cognitive function, and cancer [33].

Treatment: There are a few ways to help treat shift work disorder. One strategy is to adhere to the sleep hygiene practices that are described in Chapter 4: Sleep Hygiene Recommendations. These measures are meant to help a worker align their circadian rhythms to their new schedule and create the conditions for the best possible sleep when the sleep opportunity arises. Briefly, workers should make their sleep quarters as dark as possible, limit exposure to light in the morning, take an extended nap prior to the start of the shift, which will reduce sleepiness and potential errors during the work shift. Light therapy can be used to accelerate the resetting of the circadian clock to the new schedule. Also, melatonin may be used prior to the desired sleep schedule to help align the worker with their new circadian schedule.

Shift schedule management can also help the employees adjust to their new schedule. Workers on a rapidly rotating shift obtain less sleep than workers that are on a shift that rotates every 3 weeks. In addition, it seems easier for employees to be rotated in a clockwise rotation (moving from day to swing to night shift) than rotating in the opposite direction. Though it seems scientifically sound to keep employees on a stable night shift, evidence shows that there is only marginal improvement compared to sleep on rotating shifts. Light may be helpful in these circumstances in which bright light exposure at the beginning of the shift with an avoidance of bright light toward the end should help align the internal clock with the work schedule. Employees often get the most sleep during the swing shift, and employees may want to be sure to get sleep during this shift to dissipate much of the sleep debt that may accumulate during the other two shift rotations. During off days, it is recommended to attempt to maintain a workday schedule of sleep and wake. This can be difficult given family and social obligations. Thus one could use a compromise schedule, in which one partially reverts to the conventional schedule and attempts to avoid bright lights so as not to reentrain their circadian rhythms.

Countermeasures can also be employed during the shift to reduce sleepiness and increase alertness. Caffeine in numerous forms, including coffee, tea, and energy drinks, can be used throughout the shift. It may be best to ingest the caffeine throughout the shift so that the effects are felt throughout the shift rather than drinking a lot all at once and have the effects wane before the shift's end. In addition, naps taken before the shift can dissipate sleep debt as can naps during breaks at work. Since the onset of caffeine is 15–30 min, one suggestion is to consume caffeine and then nap for about 30 minutes before the full onset of caffeine. Other stimulants may also be used to improve wakefulness during the shift and hypnotics to help one sleep if the symptoms warrant it.

5.6   SLEEP–WAKE PHASE DISORDERS

As described in Chapter: Circadian Rhythms and Sleep-Circadian Interactions, circadian rhythms synchronize sleep and wakefulness with the Earth's day–night timing. Two sleep disorders occur when there is a mismatch between the typical day–night schedule and an individual's internal circadian rhythm. Because of the mismatch, these individuals may have difficulties falling asleep or staying asleep during sleep opportunities and unable to maintain wakefulness when necessary for work or life activities. Therefore, people that suffer from either an advance or delay of the circadian promotion of sleep and wakefulness may become sleep deprived because they have to work against their circadian biology to comply with a normal work schedule and exhibit sleepiness during periods when wakefulness is required. Other evidence suggests that the buildup and dissipation of sleep debt may be different in these people. This would mean that they may not be ready to sleep on a normal schedule. Also, the person's sensitivity to light may be altered in these people. This altered sensitivity to light may disrupt signals that signal the proper sleep time. It is best for a clinician to determine what the best course of treatment for the individual's situation.

5.6.1   Advanced sleep–wake phase disorder (ASWPD)

Definition: ASWPD occurs when a person's internal circadian clock advances faster than the typical 24-h day [36–40]. There is a coordinated cycle of proteins that keeps the body's approximately 24-h internal clock and the start of this 24-h cycle is reset daily by our first exposure to light. The amount of time it takes for proteins to be made, move into position, carry out their required actions, and then be degraded takes about 24 h allowing circadian rhythms to act as a molecular clock. The clock is synchronized to the daylight cycle so that wakefulness is promoted during daylight hours and sleep is promoted during the night. People with ASWPD, exhibit symptoms that suggest their circadian clock may be running faster than the standard circadian clock so that internally timed events occur earlier. For example in severe cases of ASWPD, an affected person will naturally go to sleep around 6 or 7 pm and then awaken around 2 am to start their day (Fig. 5.2 a and b). Based on the rest of society, these are extremely early times for either of these events.

Prevalence: The prevalence of this disorder is rare and estimated between 0.25 and ~7%. It is likely lower than delayed sleep–wake phase disorder, though the prevalence of ASWPS in the general population is dependent how strict a definition of the symptoms is applied. Another caveat to a person needs to be diagnosed to be counted. Work schedules may not be impacted if a person arrives early for work, thus people may not consider this condition to be a problem

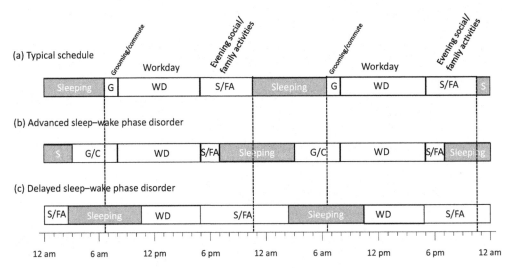

■ FIGURE 5.2 **Sleep schedules are shifted in patients with advanced or delayed sleep–wake phase disorder compared to a typical sleep schedule.** Graphs are double plotted to highlight the relationship of sleep and wakefulness over the 24-h day and how the circadian sleep disorders can be at odds with a typical schedule. (a) Graph of a typical person's schedule in which there is a consolidated sleep period from 10:30 to 6:30 each day. Grooming/commute (G/C) before work, an approximate workday (WD), and evening social/family activities (S/FA) that are typical for the day are labeled. (b) The theoretical schedule of a person with a relatively severe case of advanced sleep–wake phase disorder is displayed to highlight the effects. If one works a typical first shift, this patient would have ample morning time while the evening social time would be compressed. (c) Delayed sleep–wake phase disorder has the opposite effect, in which the habitually late bedtime makes it difficult to get a full 8 h of sleep and maintain a typical work schedule. These patients have sufficient evening time. For both circadian disorders, other individuals may have less severe cases, but the internal timing may impact the given activities.

and would not report the condition. Therefore, the actual prevalence may be hard to determine [36]. In fact, this type of schedule may benefit certain careers, such as a teacher, farmer, or west coast stock trader, where an early start is an advantage.

Risk factors: Family history is a risk factor for ASWPD. One example comes from a familial mutation in the human *Period 2* gene was found in family members that showed the symptoms of ASWPD whereas unaffected family members did not have this genetic change. The mutation in the *Period 2* gene affected the binding of casein kinase 1ε to Period 2, which alters circadian rhythms. Thus, this and other genes can be passed down through families to result in ASWPD. In addition, advancing age is a risk factor as circadian rhythms tend to shift earlier in older adults.

Consequences: Many of the consequences of ASWPD are centered on nighttime activities. Evening social activities may be difficult for these individuals, as even a typical dinner party would start about their typical bedtime. Thus, any type of job that goes into the evening hours would not be ideal. Under these circumstance, the person may begin to become sleep deprived or the task may be very difficult because of building sleepiness. On the other hand, when the job timing is synchronized with their schedule, these patients tend not to lose too much sleep.

Treatment: There are no ways to change a genetic defect, thus treatments aim to align the circadian rhythms with the person's schedule using environmental and pharmacologic approaches. Bright light therapy in the evening and melatonin in the morning are meant to delay bedtime and rise times to get them on a more conventional schedule. In addition hypnotics may be tried to extend sleep durations. It is unclear if these treatments work in the long term.

5.6.2 Delayed sleep–wake phase disorder (DSWPD)

Definition: DSWPD is a circadian disorder that may misalign the internal circadian clock with the day–light schedule [37–42]. DSWPD results in a delayed bedtime and rise time, the opposite effect of ASWPD. In severe cases, the person will not fall asleep until between 2 and 6 am. The person will then rise about 10 am to 2 pm for 8 h of sleep (Fig. 5.2 A and C). A typical 11 pm bedtime falls during a period called the wake maintenance zone for a person with DSWPD. During this period, circadian rhythms are signaling wake and resisting sleep, which makes it nearly impossible to fall asleep at this time. To make a typical work start time, the person is forced to wake up earlier than the time when they would naturally do so, building sleep debt. Moreover, wake time then occurs during a time when the circadian system is promoting sleep. Thus, individuals with DSWPD are likely to be sleep deprived on a typical schedule.

Prevalence: As with ASWPD, it has been hard to determine a widespread prevalence. Prevalence has been estimated at between 0.2 and 10% of the general population depending on the definition applied, but prevalence is likely higher in adolescents and people with insomnia.

Consequences: There are some medical conditions associated with DSWPD, including depression and the use of antacids, hypnotics, tobacco, alcohol, and caffeine. Patients with DSWPD may have a hard time adapting to a normal work schedule. This condition is like having perpetual jet lag from traveling west to east, which is a more difficult adaptation. The consequences then lead to sleepiness, irritability, a lack of concentration, and increased job loss.

Risk factors: As with other circadian disorders, numerous proteins involved in the core circadian clock that are associated with DSWPD. These include the human *Period 1*, *Period 2*, *Period 3*, and *Clock* gene as well as other genes that can alter the circadian clock. The hypothesis is that the genetic alteration slows the circadian clock leading to a delayed bedtime and rise time. Therefore, DSWPD can run in families as variants of these genes are passed from generation to generation.

Treatment: Given that this syndrome is likely the result of a misaligned circadian clock, there are therapies that can begin to address the problem. Bright light therapy can be used in the morning and a reduction to light in the evening may help phase advance the circadian clock. Evening melatonin can be used to begin to signal the body that it is time to sleep much earlier than the natural cycle of person. There can be different degrees of treatment to attempt to realign the circadian clock to the standard daylight schedule using light exposure and pharmaceuticals. Another strategy may be to not fight one's biology and take a job that fits their internal clock.

These sleep disorders highlight and are on the extreme ends of a continuum onto which people fall. Therefore, a savvy manager may be on the lookout for people that tend toward one chronotype (whether they are morning or night people) and try to place these people in jobs that match with their internal biology. This way, the work schedule and the internal biological clock will be more synchronized.

5.7 PARASOMNIAS

Parasomnia translates into "around sleep." According to the *International Classification of Sleep Disorders*, third edition (ICSD-3), parasomnias are (1) recurrent episodes of incomplete awakening, (2) absent or inappropriate responsiveness, (3) limited or no cognition or dream report, and (4) partial or complete amnesia for the episode. The parasomnias fall into two categories depending on which stage of sleep they occur. Conditions such as sleep walking, sleep terrors, sleep-related eating, sleep violence, and sleep sex occur during and around the deepest form of sleep, slow wave sleep (SWS), whereas REM behavior disorder occurs during REM sleep. These conditions hint at the fact that sleep is not a unitary process because these people exhibit brain waves that indicate that they are both asleep and carry out goal directed behaviors as one would when they are awake. Supporting this hypothesis, the person is not properly responsive to external stimuli and is not able to understand the nature and consequences of their actions. Since this is atypical behavior and goes against our understanding of what sleep is, it can be hard to fathom that a person is both asleep and carrying out the actions that they are. In some extreme examples of this type of behavior, people have gotten up and carried out violent acts against others and then are completely unaware and cannot remember the actions they took.

5.7.1 Non-REM parasomnias: sleepwalking, sleep-related eating, sleep violence, sleep sex, night terrors

Sleepwalking definition: Sleepwalking is also called somnambulism and is probably the most well known of the parasomnias [43–46]. In its most animated form, it occurs when a person is asleep and bolts upright, gets out of bed, and begins walking or possibly running in a motivated or possibly in an escape-like manner. In this state, the person can carry out automatic actions that do not require higher-level cognition such as memory, mental integration, planning, or social engagement. A sleepwalking individual is aware enough of their surroundings to negotiate a path toward a destination and can achieve their goals often carrying out an instinctual or automatic behaviors such as getting dressed, sleep-related eating, violence, or sexual activity. Episode generally lasts between 1 and 15 min. Episodes may occur from about once a week or so to once a night or more. As with most diseases, there is a spectrum of severity from people who bolt up in bed and exhibit confusing arousals to those that run and vocalize in the sleepwalking state.

Prior to the development of the EEG, there was the misconception that sleepwalking was associated with a response to dreams. Once sleepwalking was examined in conjunction with brain-wave analysis, it was found that sleepwalking occurred during SWS rather than REM sleep. Therefore, it was unlikely that people were responding to the fantastical and emotional dreams that one associates with such dramatic actions, since SWS is incompatible with the disengagement that occurs with our typical dream state. Although some sleepwalkers will claim that the movement is in response to a threatening situation or dream, researchers have hypothesized that sleepwalkers recall dreams from REM sleep that occurs after the sleepwalking event not during sleepwalking.

Sleepwalkers will report that they have a sense of anxiety during the episode and that they are "half aware" of their surroundings. Further study and observations have led to the hypothesis that sleepwalking is a disease of arousal. Under this hypothesis, much of the brain is in the deepest most disengaged form of sleep. During this period, part of the brain shows signs of arousal thus permitting these people to both move throughout the room as well as be unresponsive to external stimuli. In fact, EEG recordings show that sleepwalkers have more SWS than controls but they have more interruptions and microarousals during the SWS period. As discussed subsequently, conditions that enhance SWS tend to increase the likelihood of a sleepwalking event. In fact, one group has proposed that one can induce sleepwalking events through a protocol of sleep deprivation, which will enhance SWS and during the SWS they initiate auditory stimulation that will cause arousal in the person. This combines the two critical factors that seem to be major features of sleepwalking, arousal during SWS. Because SWS predominates in the first one-third of the night, sleepwalking tends to occur during this time of night.

There are other automatic behaviors that occur either in conjunction with sleepwalking or confusional arousals:

Sleep-related eating: This parasomnia involves sleepwalking, often to the kitchen, to consume high-caloric foods, perform potentially dangerous food preparation, and sometimes consume nonfood items. Injuries such as drinking hot liquids, choking, lacerations, consuming things such as buttered cigarettes or glue, or items they are allergic to. There is no memory of the event in the morning. Onset of sleep-related eating is around age 24 and occurs primarily in women. Sleep-related eating can occur multiple times a night and over multiple nights a week. Therefore, there are also metabolic changes, such as excessive weight gain and associated metabolic changes and potential depressive consequences that result from overeating [47].

Sleep terror: This parasomnia occurs primarily in children. The child will suddenly sit upright in bed and release an ear-splitting and alarming scream, which is accompanied by symptoms consistent with activation of the "flight or fight" portion of the nervous system. The child is often unresponsive and inconsolable. The event is over in a few minutes and the child will go back to sleep with little recall of the event upon awakening. This occurrence is often more traumatic for the witness than the patient [44].

Sleep violence: There have been reports throughout the literature of extremely violent acts carried out by individuals that show all the hallmarks of being asleep. In one case, a vacationing detective helped local police catch a murderer. The murder was carried out by the detective while he had sleepwalked. He had no recollection of the event. Sleep violence is more prevalent with young males. Overall, about 1.7% of people report sleep violence associated with sleep walking and sleep terrors, and this form of violence toward nearby people is more frequent than REM behavior disorder [48].

Sexsomnia: This action is classified as a confusional arousal including prolonged or violent masturbation, sexual molestation and assaults, initiation of sexual intercourse, and loud sexual vocalizations [49]. These acts are followed by morning amnesia of the events. This is reported to occur primarily in men.

Prevalence: Sleepwalking is most common in children but can also occur in adults. It is difficult to get a true prevalence because one often has to rely on people going to see a sleep physician or wait for an injury to occur to report a sleepwalking event, which leads to inaccurate sampling. To assess prevalence, investigators have begun to reach out into communities to survey families. The caveat to this approach is the scientists are relying on familial self-report and the family may not be aware or may not be able to recognize subtle differences in phenotypes. But this approach does begin to give a sense of how prevalent sleep walking is. Some general findings are that parasomnias are found in 1–5% of the population with sleepwalking being a slightly lower percentage and confusional arousals being the most prevalent. A few studies have found a gender difference, but these findings are not consistent. It has been reported that between 0.8 and 17% of children sleepwalk with the peak being around age 11 or 12. With age, the prevalence of parasomnias decreases. About 18–24% of sleepwalking children will continue to sleepwalk into adulthood, but if sleepwalking persists into adulthood, it is highly likely the person was a sleepwalking child. Rarely, does sleepwalking appear strictly in adulthood.

Risk factors: The risk factors for sleepwalking are aspects that enhance the combination of deep SWS with arousals. There are a number of ways to increase the intensity of SWS or create conditions in which the brain maintains a sleep-like state. Sleep deprivation is a common risk factor because it increases SWS during the recovery period, which could contribute to part of the brain remaining asleep during an arousal. Other mechanisms that may contribute to the risk of sleepwalking include alcohol intake and the use of sleeping pills. Though the evidence is more anecdotal, both of these substances depress the nervous system such that the brain will remain in a sleep-like disengagement during an arousal event. On the other hand, conditions that increase arousal during the typical SWS also increase sleepwalking events. In one study, sleepwalking children had a high prevalence of other sleep disorders, including sleep apnea and restless leg syndrome. These conditions likely provide the arousal stimulus during SWS that leads to sleepwalking. In fact, if the accompanying sleep disorder is treated, then the parasomnias decrease. Also, children with attention deficit hyperarousal disorder are also at increased risk for sleepwalking, possibly because of the increased arousals that may occur. Thus, people with sleepwalking may be good candidates for evaluation at a sleep lab to confirm that there are not comorbid sleep disorders. When people sleep away from home, sleep tends to be more restless and more fragmented by arousals. As could be predicted, sleepwalking occurs more frequently when people sleep in a strange bed.

Anecdotal evidence suggests that sleepwalking runs in families. There is some support that there is a genetic component to this condition. Often upon taking medical histories, sleepwalking is present in close relatives. In addition, one family was studied using genetic techniques that identify specific genes that are associated with the condition. This study identified a particular area on chromosome 20 that tracked with the disease. The authors suggested an autosomal dominant relationship with variable penetrance. On the contrary, twin studies did not show any special association with sleepwalking. It is likely that there will be genes associated with sleepwalking, but that there will be different causes derived from different genetic mutations.

Consequences: The primary consequence of sleepwalking alone is the risk of injury to the sleepwalker. Typically, sleep walking may be accompanied by other parasomnias, such as sleep sex or sleep violence, in which others may be injured or traumatized by the actions carried out during sleep and sleep-related eating in which the individual may try to cook and unwittingly hurt themselves or others through the use of the stove or oven, which can burn or start a fire. In fact, sleepwalkers exhibit a high proportion of injuries to themselves or others with 29 out of 41 sleepwalkers reporting these types of injuries. In the case of sleep-related eating, individuals will put on weight by unwittingly eating more calories than necessary.

Treatment: There are ways to decrease the incidence of sleepwalking and other parasomnias. A typically effective treatment of parasomnia conditions is Clonezapam or other benzodiazipines. There are reports that antidepressant drugs has a positive effect in at least a subpopulation of people with parasomnias. Less traditional therapies, such as hypnosis, have also been described as effective, but this is more anecdotal and has not been subjected to rigorous study.

5.7.2 **REM behavior disorder**

Definition: As the name states, REM behavior disorder occurs during rapid eye movement (REM) sleep [50–54]. REM is the stage of sleep in which we have our more fantastical and emotional dreams. Sometimes these dreams are threatening to us or our loved ones or they may be frightening in which a person needs to escape. To prevent an individual from acting out during our dreams, the body inactivates the neurons that permit us to carry out voluntary movement. Thus, any potentially dangerous action that we might try to carry out in response to the fantasy portions of the dream is ineffective. One of the only muscles that we can move are our eyes, which the darting eyes gives this sleep stage its name.

In REM behavior disorder, the mechanism responsible for inhibiting voluntary motor activity no longer functions properly. Therefore, individuals are then able to act in response to their dreams. The content of the dreams usually consists of the person or a loved one being attacked or under threat from an unfamiliar person or a threatening animal. In response, the person claims to be forcibly defending or escaping the threat. These responses range from vocalizations to violent behaviors. There are documented cases of where husbands have beaten or strangled their bed partners in response to what they thought was a threatening situation in which someone they cared about was in trouble. The husbands are often described as a peaceful, loving person and there had been no history of abuse. Those that study this particular affliction consistently reiterate that this is not a manifestation of latent violent tendencies. It is the inappropriate manifestation of the mechanisms associated with REM sleep. The husband often does not realize what he is doing and does not remember the physical aspects of the previous night. In other instances, people have jumped into closet doors or out windows to get away from a threatening dream. They have done great harm to themselves when responding to the fantasy world of dreams. Because this conditions manifests during REM sleep and REM sleep occurs primarily at the end of the night, REM behavior disorder tends to occur in the last one-third of the night.

Prevalence/risk factors: In population-based studies, the prevalence in the general population is estimated to be around 0.5%. The frequency increases with age and being male being greater risk factors for the presence of RBD. Interestingly, the incidence of RBD is much higher in patients with certain neurodegenerative disorders, including Parkinson's disease, dementia with Lewy bodies, and muscle system atrophy. Also, the use of antidepressants may increase the risk of RBD.

Consequences: There are numerous brain functions that are impaired in individuals with RBD. There are impairments in higher order processes such as executive decision making, memory, and facial expression recognition. In addition, sympathetic activity is dysfunctional, which can result in decreased heart rate variability following arousal from sleep and a lack of heart rate variability during REM sleep. Typically, heart-rate variability reflects adaptation to changes in the internal and external environments. Heart rate should be variable and under normal conditions there is considerable variability during REM sleep. The lack of heart rate variability indicates abnormalities in the autonomic system.

There appears to be a decrease in dopaminergic neurons in the brains of individuals with RBD. In fact, RBD is associated with various neurologic disorders, especially with neurodegenerative diseases known as synucleinopathy, including Parkinson's disease, dementia with Lewy bodies (LBD), and multiple system atrophy (MSA). Parkinson's disease is, at least in part, a disease in which dopamine signaling is decreased. Therefore, the reduction in dopaminergic signaling may manifest itself first as RBD and then eventually as Parkinson's disease. On average, the onset of Parkinson's disease is 12.7 years after the onset of RBD.

Treatment: Treatment should be sought as soon as symptoms consistent with RBD are noticed by the bed partner, especially if there is injury to either person. People suffering from RBD should be reassured that they are not mentally ill. On the other hand, there are steps that can be taken that prepare for a better outcome. There are restraints that can help with the uncontrolled outbursts. In addition, partners in a couple can sleep separately, furniture can be removed from around the bed, and the mattress can be moved to the floor. These measures can help prevent injury to either person during an RBD episode.

If these measures are inadequate or unacceptable, clonazepam is effective at reducing the frequency and severity of the behaviors in approximately 90% of patients as well as showing a decrease in unpleasant dream recall. In addition, melatonin has been shown to reduce the number of REM sleep with activity episodes during the night. Other medications, such as pramipexole, levodopa, paroxetine, and acetylcholinesterase inhibitors (donepezil and rivastigmine) are also treatment options if others fail. Given the severe consequences of RBD for family members, treatment is essential and should be geared toward the individual. Given the likely age of these people, medications that are used should take into consideration cognitive impairment and falls so that the medications do not create further problems.

5.8 FATAL FAMILIAL INSOMNIA

Definition: The hallmark of fatal familial insomnia (FFI) is the untreatable insomnia that manifests rapidly beginning at around age 50 [55–60]. Within months of disease onset, there is little detectable slow wave sleep (SWS) or REM sleep. Patients will not initiate SWS and REM in the usual cycle, instead they will initiate these sleep stages inappropriately from wakefulness. As the disease progresses, sleep times progressively decrease and patients will have persistent drowsiness with periods of voluntary movements that they refer to as dreams. In addition, patients exhibit speech, movement problems, and have autonomic nervous problems, including perspiration, elevated heart rate, and systemic hypertension among other symptoms. These symptoms are likely brought on by the increase in hormones associated with stress, including adrenaline and cortisol. These symptoms get progressively worse through the course of the disease.

FFI is the result of a transformation of an endogenously produced protein from one that acts normally to one that aggregates within neurons. This protein is the prion protein and forms of it are responsible for both FFI as well as other spongiform encephalopathies, such as Creutzfeldt–Jakob disease, which is akin to the human form of mad cow disease. In the FFI version of the prion disease, a mutation to the DNA converts an aspartate to as asparagine at position 178 in the protein. This in conjunction with either one copy (heterozygous) or two copies (homozygous) of a methionine at position 129, dictates the severity of the disease. The disease progression is more severe with two copies of the methionine. Under normal circumstances, prion protein does not build up in the brain. In FFI, the protein aggregates and remains in the brain, which causes the cells of the brain to work less efficiently and many of the cells, including in critical parts of the brain described subsequently. Since the mutation that leads to FFI is encoded in the genome, it can be potentially passed from the affected parent to their offspring and thus handed down in families. As with other prion diseases, the mutated FFI prion protein can be injected into the brain of a healthy animal and a few months later, the injected animal will show the hallmarks of FFI. Despite the potential for transmissibility, it is unclear if there have been any cases of human-to-human transfer.

The aggregates of the prion protein injure specific parts of the brain. Even when disease progression is quick, parts of the thalamus are degraded while much of the rest of the brain does not exhibit the same extent of damage. The thalamus is a region of the brain that sorts sensory information and communicates with the higher processing centers, including emotional areas, of the brain. Thus, the thalamus is a critical portion of the brain. Because of the symptoms seen in individuals with FFI, it adds more evidence to the role that the thalamus plays in sleep regulation.

Prevalence: FFI is a rare disease and it was estimated that only about 25 families spread over the globe that have been diagnosed with FFI. These are likely sporadic cases as they are unrelated families. As an understanding of the symptoms is disseminated to sleep practitioners throughout the world are aware of the condition and new families continue to be identified.

Risk factors: FFI is a consequence of a random or sporadic mutation that alters a particular amino acid in the prion protein. There are no specific risk factors or populations that are more vulnerable. Therefore, the only way to reduce the risk for FFI is to reduce the risk of random mutations by avoiding substances or conditions that induce mutations in the DNA, such as carcinogens.

Consequences: For individuals with two copies of the methionine at codon 129, it takes about 12 months between onset of symptoms and death. For a person with only one methionine, it takes around 21 months between onset and demise.

Treatment: There is currently no treatment option for individuals with FFI. Traditional sleep remedies and pharmaceuticals do not treat the sleeplessness.

5.9 HYPERSOMNIAS

Definition: Hypersomnias describe a set of conditions that results in excessive sleep or drive to go to sleep during the primary wake period [61–63]. The person may not be able to sustain wakefulness or alertness when it is required. Often the person will have involuntary sleep episodes while watching TV, reading, during social events, having conversations, or while driving. Nearly every sleep disorder listed in this chapter can lead to hypersomnia under the right circumstances. In particular, sleep apnea, narcolepsy, and restless leg syndrome can lead to hypersomnia because of the shortened or fragmented primary sleep period. In addition, behaviorally induced sleep syndrome (BISS), or sacrificing sleep for either social, relaxation or work responsibilities, is another source of EDS observed in workers. Since these topics have been covered earlier in this chapter and in Chapter: Basics of Sleep Biology, we will only introduce the novel ideas here, including idiopathic hypersomnia. Idiopathic hypersomnia is a condition in which the individual has constant EDS despite adequate sleep time. There may be the urge to take three to four naps per day, and there may be some sleep inertia felt upon waking. What makes idiopathic hypersomnia so vexing is that there is no discernable cause to the hypersomnia.

Prevalence: Idiopathic hypersomnia has been estimated to affect 0.005% of the population. There does not appear to have a gender difference. The age of onset is somewhere between the 20s and 40s.

Risk factors: A genetic predisposition is suggested, with a possible autosomal-dominant pattern of inheritance. Although narcolepsy is associated with specific HLA haplotypes, recent studies have not found any association between idiopathic hypersomnia and specific HLA markers. There are numerous medications (Textbox 1) and medical conditions (Textbox 2) that may contribute to or result in hypersomnia.

TEXTBOX 1   EXAMPLES OF MEDICATIONS THAT MAY CAUSE HYPERSOMNIA

- Benzodiazepine receptor agonists (BZRAs)
- Melatonin receptor agonists
- Histamine H1 antagonists
- Sleeping pills
- Tricyclic antidepressants
- Anxiolytics
- Antihistamines
- Antiepileptics
- Antihypertensives

TEXTBOX 2   MEDICAL REASONS FOR HYPERSOMNIA

- Metabolic or endocrine
  - Hypothyroidism
  - Cushing syndrome
  - Menopause
  - Diabetes

- Obesity
- Depression
- Infectious disorders
  - HIV
  - Guillain–Barré syndrome
  - Infectious mononucleosis
  - Lyme disease
  - Meningitis
- Parkinson's or Alzheimer's disease
- Stroke
- Tumors: thalamus, hypothalamus, brainstem, pituitary
- Multiple sclerosis
  - The treatment, interferon, can lead to hypersomnia
- Genetic disorders
  - Niemann–Pick type C
  - Prader–Willi syndrome

Consequences: The consequences lie in managing the unrelenting sleepiness. This can lead to inadvertently falling asleep and to workplace accidents or accidents in public. Hypersomnia is associated with many of the health problems described with sleep loss and is associated with a shorter lifespan.

Treatment: Nonpharmacologic management of symptoms may be successful. These measures include adhering to sleep hygiene principles and avoiding sleep deprivation. If these behavioral modifications are unsuccessful, stimulant medications can be used to promote wakefulness. These compounds include methamphetamine, methylphenidate, modafinil, armodafinil, or sodium oxybate. This will help maintain consolidated wakefulness and importantly alertness.

Sleep disorders can disrupt many different mechanisms that govern sleep and wakefulness. This will result in a shorter sleep duration and more fragmented sleep, which often results in EDS. The increased sleepiness is also accompanied by an increase in depressive symptoms. Though sleep disorders seem to be very pervasive, there are treatments and behavioral recommendations to manage the impact the sleep disorder may have on the individual's life.

REFERENCES

[1] Stansbury RC, Strollo PJ. Clinical manifestations of sleep apnea. J Thorac Dis 2015;7(9):E298–E2310.

[2] Wellman A, White DP. Central sleep apnea and periodic breathing. In: Kryger MH, Roth MH, Dement WC, editors. Principles and practice of sleep medicine. Saint Louis, MO: Elsevier Saunders; 2011. p. 1140–52.

[3] Weaver TE, George CFP. Cognition and performance in patients with obstructive sleep apnea. In: Kryger MH, Roth T, Dement WC, editors. Principles and practice of sleep medicine. Saint Louis, MO: Elsevier Saunders; 2011. p. 1194–205.

[4] Cao TC, Guillenminault C, Kushida CA. Clinical features and evaluation of obstructive sleep apnea and upper airway resistance syndrome. In: Kryger MH, Roth T, Dement WC, editors. Principles and practice of sleep medicine. Saint Louis, MO: Elsevier Saunders; 2011. p. 1206–18.

[5] Buchanan P, Grunstein R. Positive airway pressure treatment for obstructive sleep apnea-hypopnea syndrome. In: Kryger MH, Roth T, Dement WC, editors. Principles and practice of sleep medicine. Saint Louis, MO: Elsevier Saunders; 2011. p. 1233–49.

[6] Phillips BA, Kryger MH. Management of sleep apnea-hypopnea syndrome. In: Kryger MH, Roth T, Dement WC, editors. Principles and practice of sleep medicine. Saint Louis, MO: Elsevier Saunders; 2011. p. 1278–93.

[7] Polotsky VY, Jun J, Punjabi NM. Obstructive sleep apnea and metabolic dysfunction. In: Kryger MH, Roth T, Dement WC, editors. Principles and practice of sleep medicine. Saint Louis, MO: Elsevier Saunders; 2011. p. 1331–8.

[8] Somers VK, Javaheri S. Cardiovascular effects of sleep-related breathing disorders. In: Kryger MH, Roth T, Dement WC, editors. Principles and practice of sleep medicine. Saint Louis, MO: Elsevier Saunders; 2011. p. 1370–80.

[9] Young T, Nieto FJ, Javaheri S. Systemic and pulmonary hypertension in ostructive sleep apnea. In: Kryger MH, Roth T, Dement WC, editors. Principles and Practice of Sleep Medicine. Saint Louis, MO: Elsevier Saunders; 2011. p. 1381–92.

[10] Hedner J, Franklin KA, Peker Y. Coronary artery disease and obstructive sleep apnea. In: Kryger MH, Roth T, Dement WC, editors. Principles and Practice of Sleep Medicine. Saint Louis, MO: Elsevier Saunders; 2011. p. 1393–9.

[11] Franklin KA, Lindberg E. Obstructive sleep apnea is a common disorder in the population-a review on the epidemiology of sleep apnea. J Thorac Dis 2015;7(8):1311–22.

[12] Bratton DJ, et al. CPAP vs mandibular advancement devices and blood pressure in patients with obstructive sleep apnea: a systematic review and meta-analysis. JAMA 2015;314(21):2280–93.

[13] Boethel CD, Al-Sadi A, Barker SB. Residual sleepiness in obstructive sleep apnea: differential diagnosis, evaluation, and possible causes. Sleep Med Clin 2013;8(4):571–82.

[14] Eckert DJ, et al. Central sleep apnea: pathophysiology and treatment. Chest 2007;131(2):595–607.

[15] Vgontzas AN, Fernandez-Mendoza J. Insomnia with short sleep duration: nosologic, diagnostic, and treatment implications. Sleep Med Clin 2013;8(3):309–22.

[16] Morin CM, Jarrin DC. Epidemiology of insomnia: prevalence, course, risk factors, and public health burden. Sleep Med Clin 2013;8(3):281–97.

[17] Pigeon WR, Cribbet MR. The pathophysiology of insomnia: from models to molecules (and back). Curr Opin Pulm Med 2012;18(6):546–53.

[18] Lichstein KL, et al. Insomnia: epidemiology and risk factors. In: Kryger MH, Roth T, Dement WC, editors. Principles and practice of sleep medicine. Saint Louis, MO: Elsevier Saunders; 2011. p. 827–37.

[19] Harvey AG, Spielman AJ. Insomnia: diagnosis, assessment, and outcomes. In: Kryger MH, Roth T, Dement WC, editors. Principles and practice of sleep medicine. Saint Louis, MO: Elsevier Saunders; 2011. p. 838–49.

[20] Perlis M, et al. Models of insomnia. In: Kryger MH, Roth T, Dement WC, editors. Principles and practice of sleep medicine. Saint Louis, MO: Elsevier Saunders; 2011. p. 850–65.

[21] Roth T. Insomnia: definition, prevalence, etiology, and consequences. J Clin Sleep Med 2007;3(5 Suppl):S7–10.

[22] Overeem S, Reading P, Bassetti CL. Narcolepsy. Sleep Med Clin 2012;7(2):263–81.

[23] Faraco J, Mignot E. Genetics of narcolepsy. Sleep Med Clin 2011;6(2):217–28.

[24] Mignot E. Narcolepsy: pathophysiology and genetic predisposition. In: Kryger MH, Roth T, Dement WC, editors. Principles and practice of sleep medicine. Saint Louis, MO: Elsevier Saunders; 2011. p. 938–56.

[25] Guilleminault C, Cao MT. Narcolepsy: diagnosis and management. In: Kryger MH, Roth T, Dement WC, editors. Principles and practice of sleep medicine. Saint Louis, MO: Elsevier Saunders; 2011. p. 957–68.

[26] Thannickal TC, et al. Reduced number of hypocretin neurons in human narcolepsy. Neuron 2000;27(3):469–74.

[27] Chemelli RM, et al. Narcolepsy in orexin knockout mice: molecular genetics of sleep regulation. Cell 1999;98(4):437–51.

[28] Imamura S, Kushida CA. Restless legs. Sleep Med Clin 2014;9(4):513–21.

[29] Stevens MS. Restless legs syndrome/Willis–Ekbom disease morbidity: quality of life, cardiovascular aspects, and sleep. Sleep Med Clin 2015;10(3):369–73.

[30] Montplaisir J, et al. Restless legs syndrome and periodic limb movements during sleep. In: Kryger MH, Roth T, Dement WC, editors. Principles and practices of sleep medicine. Saint Louis, MO: Elsevier Saunders; 2011. p. 1026–37.

[31] Stefansson H, et al. A genetic risk factor for periodic limb movements in sleep. N Engl J Med 2007;357(7):639–47.

[32] Akerstedt T, Wright KP Jr. Sleep loss and fatigue in shift work and shift work disorder. Sleep Med Clin 2009;4(2):257–71.

[33] Reid KJ, Abbott SM. Jet lag and shift work disorder. Sleep Med Clin 2015;10(4):523–35.

[34] Drake CL, Wright KP. Shift Work, Shift-Work Disorder, and Jet Lag. In: Kryger MH, Roth T, Dement WC, editors. Principles and practices of sleep medicine. Saint Louis, MO: Elsevier Saunders; 2011. p. 784–98.

[35] Drake CL, et al. Shift work sleep disorder: prevalence and consequences beyond that of symptomatic day workers. Sleep 2004;27(8):1453–62.

[36] Auger RR. Advance-related sleep complaints and advance sleep phase disorder. Sleep Med Clin 2009;4(2).

[37] Dodson ER, Zee PC. Therapeutics for circadian rhythm sleep disorders. Sleep Med Clin 2010;5(4):701–15.

[38] Reid KJ, Zee PC. Circadian disorders of the sleep–wake cycle. In: Kryger MH, Roth T, Dement WC, editors. Principles and practice of sleep medicine. Saint Louis, MO: Elsevier Saunders; 2011. p. 470–82.

[39] Kim MJ, Lee JH, Duffy JF. Circadian rhythm sleep disorders. J Clin Outcomes Manag 2013;20(11):513–28.

[40] Toh KL, et al. An hPer2 phosphorylation site mutation in familial advanced sleep phase syndrome. Science 2001;291(5506):1040–3.

[41] Lack LC, Wright HR, Bootzin RR. Delayed sleep-phase disorder. Sleep Med Clin 2009;4(2):229–39.

[42] Micic G, et al. The etiology of delayed sleep phase disorder. Sleep Med Rev 2015;27:29–38.

[43] Banerjee D, Nisbet A. Sleepwalking. Sleep Med Clin 2011;6(4):401–16.

[44] Modi RR, Camacho M, Valerio J. Confusional arousals, sleep terrors, and sleepwalking. Sleep Med Clin 2014;9(4):537–51.

[45] Mahowald MW, Cramer Bornemann MA. Non-REM arousal parasomnias. In: Kryger MH, Roth T, Dement WC, editors. Principles and practices of sleep medicine. Saint Louis, MO: Elsevier Saunders; 2011. p. 1075–82.

[46] Zadra A, et al. Somnambulism: clinical aspects and pathophysiological hypotheses. Lancet Neurol 2013;12(3):285–94.

[47] Howell MJ. Sleep eating. Sleep Med Clin 2011;6(4):429–39.

[48] Ohayon MM, Schenck CH. Violent behavior during sleep: prevalence, comorbidity and consequences. Sleep Med 2010;11(9):941–6.

[49] Buchanan PR. Sleep sex. Sleep Med Clin 2011;6(4):417–28.

[50] Tachibana N. REM sleep behavior disorder. Sleep Med Clin 2011;6(4):459–68.

[51] Howell MJ, Schenck CH. Rapid eye movement sleep behavior disorder and neurodegenerative disease. JAMA Neurol 2015;72(6):707–12.

[52] Mahowald MW, Schenck CH. REM sleep parasomnias. In: Kryger MH, Roth T, Dement WC, editors. Principles and practices of sleep medicine. Saint Louis, MO: Elsevier Saunders; 2011. p. 1083–97.

[53] Peever J, Luppi PH, Montplaisir J. Breakdown in REM sleep circuitry underlies REM sleep behavior disorder. Trends Neurosci 2014;37(5):279–88.

[54] Schenck CH, Mahowald MW. REM sleep behavior disorder: clinical, developmental, and neuroscience perspectives 16 years after its formal identification in SLEEP. Sleep 2002;25(2):120–38.

[55] Gambetti P, Lugaresi E. Conclusions of the symposium. Brain Pathol 1998;8(3): 571–5.

[56] Lugaresi E, et al. Fatal familial insomnia and dysautonomia with selective degeneration of thalamic nuclei. N Engl J Med 1986;315(16):997–1003.

[57] Lugaresi E, et al. The pathophysiology of fatal familial insomnia. Brain Pathol 1998;8(3):521–6.

[58] Medori R, et al. Fatal familial insomnia: a second kindred with mutation of prion protein gene at codon 178. Neurology 1992;42(3 Pt 1):669–70.

[59] Montagna P, et al. Clinical features of fatal familial insomnia: phenotypic variability in relation to a polymorphism at codon 129 of the prion protein gene. Brain Pathol 1998;8(3):515–20.

[60] Goldfarb LG, et al. Fatal familial insomnia and familial Creutzfeldt–Jakob disease: disease phenotype determined by a DNA polymorphism. Science 1992;258(5083):806–8.

[61] Ahmed I, Harris S, Thorpy MJ. Differential diagnosis of hypersomnias. Sleep Med Clin 2012;7(2):191–204.

[62] Bassetti CL, Dauvilliers Y. Idiopathic hypersomnia. In: Kryger MH, Roth T, Dement WC, editors. Principles and practices of sleep medicine. Saint Louis, MO: Elsevier Saunders; 2011. p. 969–79.

[63] Billiard M, Sonka K. Idiopathic hypersomnia. Sleep Med Rev 2015;29:23–33.

Chapter 6

Fatigue and human performance

The most obvious effect of sleep deprivation is sleepiness which is characterized by fatigue, lack of motivation, and the urge to nod off. Scientists measure sleepiness in a variety of ways including brain activity. Using an electroencephalogram (EEG) they have found differences in brain activity when someone is alert, sleepy, or asleep. These changes correspond to a lower level of cognitive performance and a susceptibility to sleep. Generally being awake for a more than the typical 16 h will lead to measurable biological changes and associated declines in performance. This can include drops in concentration, working memory, analytical capacity, and reasoning ability.

The functioning of the human brain is intricate and the effect of sleep deprivation is complex. There is also significant variation between individuals and there can be noticeable difference for the same individual at different times. Many areas of the brain exhibit changes with sleep deprivation, but one prominent area in which changes in activity level correlate with fatigue-related decline is the region known as the prefrontal cortex (PFC). It is responsible for many of our higher-level cognitive functions. This portion of the brain is particularly vulnerable to a lack of sleep. This causes sleep-deprived people to show deficits in many tasks that require logical reasoning or complex thought. This combined with other biological processes will cause meaningful deterioration in human performance and potential workplace safety as people become fatigued.

6.1 FATIGUE AND HUMAN ERROR

The medical profession is an example of one group that has become concerned about connection between errors and sleep deprivation. A study in the *New England Journal of Medicine* explored sleep deprivation among medical interns and the occurrence of serious medical errors. This comprehensive study [1] compares errors while interns either followed a traditional work schedule or one that was designed to reduce sleep deprivation. The study took place at Boston's Brigham and Women's Hospital's medical intensive care unit (MICU), and its coronary care unit (CCU). Fig. 6.1

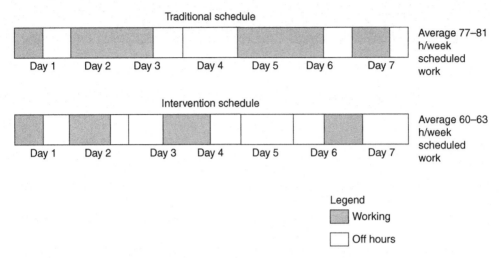

■ FIGURE 6.1 Two work schedules used in study of work hours and errors [1].

illustrates the two work schedules that were used in the study. The traditional schedule consists of a daytime swing shift followed by an extended on-call shift and a day off. The interns' averaged 77–81 h working with up to 34 continuous hours in a workweek with this schedule. The intervention schedule included a standard swing shift and a day call shift the following day. Then the interns worked night-call shifts. This modified schedule resulted in 60–63 h work weeks and consecutive hours were limited to about 16 h. Due to the nature of healthcare the interns often exceeded the schedules shown in Fig. 6.1. In practice, the intervention eliminated working more than 24 continuous hours and reduced the overall work by approximately 20 h a week and increased the intern's sleep by nearly 1 h a day.

The study's results are striking. Interns made nearly 36% more serious medical errors during the traditional schedule than during the intervention schedule. Serious diagnostic errors were 5.6 times higher during the traditional schedule compared with the intervention schedule. The authors concluded that working fewer hours reduces the number of serious medical errors. Interns, like the rest of us, perform better when we are not sleep deprived and overly fatigued.

In response to this study and others, the Accreditation Council for Graduate Medical Education (ACGME) established requirements beginning in 2003 to limit the hours medical residents can work to promote safe healthcare. Journal papers have been written exploring the results of the ACGME restriction on duty hours for medical residents. Fletcher, Reed, and Arora [2] conducted a systematic review of over 4800 journal papers. They found the impact on restricted duty hours was positive. The number of medical errors decreased and the residents' health improved. Their specific findings related to limiting working hours were the following:

- Reduced working hours for medical residents reduced the number of patient death.
- Working fewer hours did not affect the residents standardize test scores but they did receive fewer hours of operative experience.
- Burnout among those studying to be doctors was decreased and their well-being improved.
- The potential for workload to affect the rate of patient medical complications needs more study.

6.2 **FATIGUE AND HAND-EYE COORDINATION**

It is generally accepted that people do not perform as well when they are fatigued as they do when they are well rested. As is often the case in human factors, the question becomes how do we quantify something related to humans? In recent years sleep researchers have found that comparing an individual's performance while sleepy to that of being drunk is a useful comparison [3]. The effects of alcohol on human performance has become an acceptable standard for safe performance. Few would argue that a person with an elevated blood alcohol concentration (BAC) would be safe to drive.

In a study by Dawson and Reid [4] two groups of individuals were compared. The first were kept awake for 28 h and the second drank alcohol to reach a BAC of 0.10% (legally too drunk to drive in many places). Both groups of subjects were given a cognitive psychomotor performance test that measures hand–eye coordination. As sleep deprivation and BAC increased, both groups' hand–eye coordination decreased in a similar fashion. The researchers found that 17 h of sustained wakefulness is equivalent to the impaired performance on this test of someone with a BAC of 0.05%. By the time the researcher subjects had been awake 24 h their performance decreased to a level equivalent to a BAC of roughly 0.10%.

What does that mean for industrial safety? Imagine a worker who gets up at 7:00 am after a full night's sleep. He spends the day with his family and then goes to work a night shift. By 11:00 pm, he has been up 17 h straight and his hand–eye performance is comparable to a BAC of 0.5%. By 7:00 am the next morning when his night shift was ending, he has been awake 24 h and is performing similar too someone too drunk to drive with a BAC of 0.10%. Hand–eye coordination is used when we reach and grasp things, handwrite a note, or catch a ball. It is a skill common of even the most mundane task and it deteriorates with sleep deprivation. Decreases in hand–eye coordination is just one change that happens with fatigue.

6.3 FATIGUE AND MOOD

Sleep researchers have found increased negative moods are associated with sleep deprivation in humans. This is true whether the sleep deprivation is caused by chronically not getting enough sleep at night over an extended period of time or a short term lack of any sleep. Durmer and Dinges [5] found feelings of fatigue, loss of vigor, sleepiness, and confusion to be common with sleep deprivation. Traditionally feelings of irritability, anxiety, and depression are believed to be associated with fatigue, however, experimental evidence of these mood changes due to sleep deprivation are mixed depending on the time and amount of sleep deprivation and how mood was measured. A recent study of adolescents who were deprived of one night sleep reported significant negative changes in mood [6]. The subjects reported more confusion, more anger, more fatigue, and less vigor. Additionally, significant increases in depressed mood and anxiety were reported by females. These results are consistent with adult sleep deprivation studies.

A workplace research study [7] was done to examine the relationship between sleep loss and emotional reactivity of medical residents. They studied two types of events positive goal-enhancing events (such as being able to practice a novel medical procedure) and negative goal-disruptive events (such as being interrupted while with a patient to answer a page). Perhaps not surprisingly they found that the negative events resulted in greater negative emotions when the individuals were sleep deprived. The opposite scenario was also affected. Positive events had less positive emotional response when the resident was sleep deprived compared to those who were rested. The positive emotions studied included *inspired, attentive, interested*, and *proud*. The negative emotions studied included *irritable, nervous, upset*, and *distressed*. Both sets of emotions were measured by a well-established psychological emotion scale. The takeaway from the study is that fatigued workers have amplified negative emotional response when negative things happen. When good things happen at work, fatigue damps the positive emotional response.

Another study [8] examined the relationship between sleep deprivation and feelings of frustration. The researchers used a test that involved subjects looking at cartoon-like depictions of frustrating situations. The subject said what their reply would be in the situation providing researchers insight into their feelings. The subjects that were fatigued had been awake 55 continuous hours. The researchers expected to find that sleep-deprived individuals would show greater outward expression of frustration and an increased tendency to avoid responsibility by fixing the blame on others. That is what they found. The tendency to directly blame or be outwardly hostile toward people or objects in the environment was increased for the fatigue subjects. The frequency of when subjects responded with apologizes or offers of amends to alleviate the problem were significantly lower for fatigued individuals compared to the rested control group. The findings again support increased negative emotional moods with increased fatigue.

6.4 FATIGUE AND MEMORY

Sleep researchers have considered two types of memory, declarative and procedural. Declarative is our ability to state a fact such as the capital of the country Guatemala is Guatemala City. Procedural memory is more involved. It is how we know to do some things, the steps that are required. Some have suggested that the more complex procedural memory is dependent on REM sleep. Although exact details of the relationship between memory and sleep is not clear, there is mounting evidence that human fatigue negatively affects our ability to remember information, particularly for more complex information.

Despite not fully knowing what happens cognitively during sleep, research has shown that sleep deprivation impairs two different domains of the learning and memory process. First, if one is sleep deprived prior to attempting to learn a process or procedure, the acquisition portion of the learning and memory process is impaired. Our ability to introduce new information into the brain is hampered by fatigue. The problem may be due to the reduced focus, motivation, attention, and concentration caused by fatigue and the increased sleep pressure. Ultimately, this will reduce memory because if you never acquire the information, you cannot recall it. Second, sleep helps to consolidate memories. Additional sleep helps the memory consolidation process and one is able to recall more words or additional procedures after sleep. Sleep may facilitate the process of moving information from short-term working memory to long-term memory. A simple explanation of this process is that while we are asleep the brain is able to "process," "organize," and "store" the events of the day and the things we have learned. At the same time "connections" or "information" within the brain that are not useful are "cleaned up" during sleep.

This hypothesis of what happens to the brain during sleep is visible as anatomical and molecular features in the brains of animals are detectably changed after a night of sleep. Although it is difficult to increase sleep in humans, scientists have been able to genetically increase sleep in experimental animals. The increase in sleep between the training and testing 2 days later improved memory [9]. In nap studies the increase in sleep permitted people to recall more words [10] and procedural memory than those that did not take a nap [11]. Thus, sleep is a critical part of the learning and memory process from before acquisition through consolidation and recall.

6.5 FATIGUE AND REACTION TIME

Reaction time is another human ability that is commonly evaluated in psychological research. Reaction time is the amount of time it takes to respond to a stimulus. Reactions can be defined as simple, when a subject responds with a predefined response to a defined stimulus. An example would be when the alarm sounds, push the red stop button. A choice reaction is more complex, a decision is involved in the process rather than just a single response. For example, if the driver of a car must make a decision whether to swerve or slam on the brakes in response to an emergency, this reaction will take more time. The type and characteristics of the stimuli (ie, warning lights, auditory alarms, vs displays) involved can also effect the time required to respond.

A study of simple reaction time and sleep deprivation [12] investigated how quickly performance deteriorates after 1 night without sleep when performing a monotonous task. For a 2-h period participants were required to monitor a computer screen and press a response key each time a yellow dot appeared. The dot could appear at one of four locations and at a rate of 13 times every 15 min. This task was designed to simulate a typical monitoring task that requires attention but not a high level of focus. The study found that the performance on the simple reaction time task was both significantly slower and poorer while sleep deprived. Reaction times were roughly 20% slower for people who were sleep deprived and on average they missed more than three times as many of the signals. The times became longer over the span of the test.

There are several hypotheses regarding how sleep deprivation results in reduced cognitive performance. Currently, sleep scientists are pursuing how the brain is affected by sleep deprivation using brain-imaging and psychological-testing techniques to understand how the brain acquires, processes, and determines the proper response under sleep-deprived conditions. These researchers hope to determine if there is a hierarchy to information processing and if one aspect is more vulnerable than another. In addition, one has to account for the fact that at times, certain tasks seem to be resilient to the effects of sleep deprivation, at least for a certain amount of time. A description and attempt to address this complex interaction is laid out in a meta-analysis of cognitive deficits seen with sleep deprivation. A meta-analysis is a "study of studies" in which the findings from numerous studies are combined and weighted according to the numbers of subjects that each study has. This technique was used to try and discriminate between hypotheses such as a top-down approach to information processing, a bottom-up view of information processing, and one that combines the two views.

In this study [13], they found that 24–48 h of sleep deprivation had impacts across numerous cognitive domains. The most prominent effect was the reduction of vigilance or simple attention. Though of smaller magnitude, decreases in the performance of complex attention and working memory were also found. The authors found deficits in both speed and accuracy across the tests as a whole, though each individual category did not show a significant difference. The lack of effect is, in part, due to the individual variability of the subjects based on the diverse performance of the subjects on these test, in which some people are more resilient to sleep deprivation than others. The conclusions drawn from this study are that 24–48 h of sleep deprivation takes a particularly large toll on simple attention. Although other cognitive domains are affected, this one is particularly susceptible to sleep deprivation.

These results may have a particular implication for the work environment. On the basis of these data, low stimulus–low engagement monitoring tasks would be jobs that have a high likelihood of performance errors. This is alarming because worker in these types of jobs are often the first to identify and respond to problems in industrial settings. Deficits in recognizing or adapting to the changing circumstances may have terrible consequences in detecting problems. Additionally, simple attention is a key part of reasoning and problems solving. Thus, even short stints of sleep deprivation may have major safety implications.

6.6 FATIGUE AND ATTENTION

Researchers [14] have discovered complexity in the study of how fatigue affects our ability to pay attention. Attention can be classified into intensity and selection aspects. Intensity is more fundamental and considers alertness and sustained attention. Selection considers orienting and executive attention, which could be called vigilance. Research [15] has shown that vigilance can be drastically affected by fatigue. This can be dangerous because jobs that are often associated with sleep deprivation also require vigilance; such as soldiers, pilots, or those operating complex systems.

Following are some key consequences of fatigue related to attention:

- Fatigue causes a slowing down of reaction time. This is tied to the detection of a stimuli such as a warning light or an alarm, in addition to decreases in response performance.

■ Sleep deprivation results in more errors due attention-related declines in situation awareness and vigilance. Often these are errors of omission due to failure to notice signals, but it can also be errors of commission.

■ Sleep deprivation tends to increase time requirements for an assortment task types. The fatigue-induced decline in executive functions in the human brain often manifests in difficulty focusing, concentrating, and sustaining attention to the task at hand.

Vigilance is a state where attention must be sustained over time. It is common in tasks that require some form of "keeping watch." This can be monitoring a process, observing operations, or simply listening to what is going on. When an infrequent, unpredicted, or emergency event occurs the worker often has to responds with a different action. An attention failure due to fatigue can cause error and/or delays in these situations. Humans tend to be poor at vigilance tasks. Our brains struggle to maintain sustained attention over extended periods of time. Psychologist have the term vigilance decrement to describe the decrease in human performance on vigilance tasks over time. This decline in performance can happen within 15–20 min depending on the situation.

The main factors affecting attention and our performance on vigilance tasks are:

■ Time on task: The longer we are on task, the poorer the performance.

■ Event frequency: If abnormalities are rare, human reaction times significantly increase.

■ Environment: External distractions such as noise can decrease performance on vigilance tasks, other environmental factors such as temperature can also detract from performance.

■ Sleepiness: Fatigued workers are susceptible to lapses in vigilance and an increase in response time, which may be caused by microsleeps.

■ Motivation: Both internal (ie, drive, sense of satisfaction) and external (ie, incentives, feedback) motivators play an important role in vigilance.

Changes in the monitoring task can improve a worker's vigilance performance. Table 6.1 lists examples of methods that can be used; these fall into three general categories changing operator bias, altering the system, and improving the work environment.

6.7 FATIGUE AND COGNITIVE TUNNELING

There is a concept in human factors called cognitive tunneling, inattentional blindness, or perceptual blindness. None of the names exactly capture the concept; it is not a vision problem. It is psychological; it is when a person overlooks something due to a lack of attention. When someone does not see

Table 6.1 Methods to Improve Operator Performance in Monitoring Tasks

Changing operator bias	■ Introduce "test" to change the frequency and expectation of abnormal situations ■ Provide payoffs to improve external motivation ■ Increase exhortations to improve external motivation ■ Provide feedback to influence internal motivation and awareness
Altering the system	■ Amplify and differentiate the signal to be detected (ie, louder warning alarm) ■ Make the signal dynamic (ie, flashing light) ■ Provide frequent breaks to reduce time on task and mental fatigue ■ Reduce the workload of the operator
Improving the work environment	■ Reduce distractions in the work environment such as noise ■ Provide stimulation in the environment to counter fatigue effects

a warning light or misreads a clearly written label on a chemical it may be an example of this phenomenon rather than carelessness or neglect. Unfortunately when an individual fails to recognize an unexpected stimulus that should be apparent the results can be costly. This is especially true if the message being overlooked is a warning.

At any given time our senses are taking in countless stimuli and we are not consciously aware of processing most of it. We have limited resources to focus our attention. The human brain scans about 30–40 pieces of information in the form of sights, sounds, smells, and touch per second. What captures our attention is shaped by [16]:

■ Conspicuity: Does the stimulus stand out? It could be a word that is in red or bold amid a section of normal text or hearing your name in a noise setting that causes us to focus.
■ Expectations: Does the stimulus look or sound like what we have experienced before? Our past experiences teach us what is worth focusing our attention upon. If the temperature gauge is always in the normal range we tend to not notice when it longer is normal.
■ Mental workload: Is our mind focused on a secondary task or distraction? Countless studies have shown we "miss" things when we are trying to text while driving or otherwise attempting to multitask.
■ Capacity: How many stimuli can we process? Our brain is a marvel but there are limits. Aspects of our daily lives can reduce our capacity.

■ **FIGURE 6.2** **Scene from video used in "invisible gorilla" study.** Image provided by Daniel Simons. For more information about the study and to see the video please visit www.dansimons.com or www.theinvisiblegorilla.com. Simons DJ, Chabris CF. Gorillas in our midst: sustained inattentional blindness for dynamic events perception. Perception 1999;28:1059–1074.

Fatigue has been shown to reduce our ability to process information which can effectively reduce our capacity to manage stimuli. Fatigue makes inattentional blindness more likely and the risk of human error greater.

THE INVISIBLE GORILLA STUDY

Christopher Chabris and Daniel Simons conducted the best-known study on inattentional blindness commonly referred to as the Invisible Gorilla Study. They made a video of six people passing basketballs to their teammates who were either in white shirts or black shirts. People watching the video were asked to count the number of passes made by either the white or black team. A few seconds into the video a person dressed in a black gorilla suit walks through the center of the basketball players. You would expect everyone paying enough attention to the video to count passes could not miss a gorilla. Well, of those watching the black shirted team, 17% missed the gorilla. A shocking 58% of those watching the white shirted team on the same video missed the gorilla. People were focused on the counting task and their brain did not consciously notice the gorilla. Those watching the white shirted team ignored the dark images, including the gorilla. You can look for the gorilla yourself, the video, and other related material is online at theinvisiblegorilla.com or on Youtube (Fig. 6.2).

6.8 FATIGUE AND DECISION MAKING

Decision making is a more complex mental process than simply reacting with a known response or monitoring aspects of the work environment. Sleep researchers have expected that decision making will suffer from fatigue as much or more than these simpler human activities. Researchers [17] used the Iowa Gambling Task (IGT) to test this hypothesis. The test mimics real-world decision making through the use of four decks of cards that differ in their inherent payoff schedules. Two of the decks are "good" and the subject will have lower payoff from either of these decks, but the losses are less leading to long-term gains. The remaining two decks are "bad" and the subject receives larger payout but incurs even larger penalties leading to an overall loss over the game. The test was originally used with patients who had brain lesions in the prefrontal cortex. The healthy volunteers in the control group learned to avoid the riskier "bad" decks and show a preference for the "good" decks of cards. In contrast, patients with damage to the prefrontal cortex preferred choosing from the risker "bad" decks and seldom learned that their decisions consistently lead them toward large long-term losses.

This gambling test was used with sleep-deprived volunteers to see how fatigue influenced decision making. When tested following 49.5 h of total sleep deprivation, subjects showed a decision-making pattern that became significantly worse as the game progressed. The same individuals had positive test results when they were well rested. In the sleep-deprived state they appeared less able to weigh the immediate benefits of short-term rewards against the greater costs of long-term penalties. This is a mental ability that requires integration of information, and the authors posit that when sleep deprived, the subjects were unable to attach an appropriate "good" or "bad" tag to the decks over the duration of the test. The sleep-deprived volunteers had better performance than the brain-injured patients, but their decision-making processes had striking similarities. This is in line with decreases in the brains' prefrontal cortex due to sleep deprivation. The researchers also found that older subjects tended to make riskier choices than younger subjects. Implying a negative interaction of age and fatigue on the decision-making process [17].

6.9 FATIGUE AND WORKING WITH OTHERS

Emotional intelligence (EI) is the ability to recognize emotions in both yourself and others, to distinguish different feelings, and to use this information to guide interactions with others. Emotional intelligence is key to interpersonal relationships and working well with others. A group of US Army

Researchers [18] explored the effect of sleep deprivation on this skill set by testing subjects' EI levels when rested, at 55.5 h of continuous wakefulness, and at 58 h of continuous wakefulness. They used the Bar-On Emotional Quotient Inventory (EQi) to test EI and the Constructive Thinking Inventory (CTI) to test thinking patterns. Using the EQi they found with sleep deprivation, there was a statistically significant decline in perceived emotional intelligence. This affected three major areas of functioning, including intrapersonal awareness, interpersonal skills, and stress management. Sleep deprivation was also found to produce statistically significant declines in several aspects of emotional intelligence and some aspects of constructive thinking skills, including reduced intrapersonal awareness, interpersonal functioning, stress management, and behavioral coping skills as well as an elevation in esoteric thought processes when tested via the CTI. When combined with declines in decision making and memory due to fatigue, the loss of emotional intelligence greatly hampers teams working together in complex, safety critical activities.

Interpersonal and social interactions are complex. Sleep researchers are still learning about the effects of sleep loss on every day interactions such as language interpretation. A recent study [19] focused on the impact of sleep deprivation on sarcasm detection. Sarcasm is a common pragmatic verbal tool used in many complex social functions. The use of sarcasm may make it possible to convey a criticism indirectly, getting one's message across while easing the atmosphere with humor. Accordingly, in order to respond appropriately during social interactions it is often crucial to understand whether the speaker's message should be perceived as sincere or sarcastic. The researchers compared how well people who had slept versus those who had not slept the prior night did at detecting sarcasm in a message. The ability to detect sarcasm did not decrease but the time required was increased. This is compatible with research on decreases in reaction time, decision making, and emotional intelligence due to sleep deprivation. We are still able to perform but it takes more time because we are not at our peak. The researchers [13] concluded that sleep deprivation may hamper social interactions due to a slowing in one's ability to process interactions by making it more difficult to take another person's perspective.

6.10 FATIGUE AND MARITAL LIFE

Given the varied ways human performance is negatively impacted by fatigue it is not surprising it takes a toll on our relationships outside of work as well. Obstructed sleep apnea (OSA) has been discussed in a previous chapter. Hallmarks of this sleep disorder are snoring and feelings of fatigue.

Another common occurrence is that sleeping on the same bed with a person snoring like a chainsaw due to OSA makes it hard for the bed partner to get a good night's sleep as well. Researchers studied this with a group of newly diagnosed OSA suffers [20]. One group were given CPAPs to address the sleep disorder. This device is designed to let the person with OSA have a restful night's sleep. An additional benefit is the CPAP mask stops the snoring. A second group with OSA was given a more conservative treatment that encouraged losing weight, avoiding alcohol, and sleeping a different position. As part of the study, the researchers measured the marital satisfaction for both patients and bed partners. Both patients and partners in the CPAP group reported a decrease in disagreements per week and improved marital satisfaction. Those in the non-CPAP group saw increases in disagreements and decreases in marital satisfaction. This would indicate that fatigue, whether caused by OSA or sleeping with someone who is disruptive or something else, likely will have a harmful impact on marital satisfaction.

A similar result was found in a study of middle-aged women [21]. Happily married women reported fewer sleep disturbances, including less difficulty falling asleep, fewer night awakenings, fewer early morning awakenings, and less restless sleep, as compared to women reporting lower marital happiness. This result held true when factors that have shown previous associations with sleep (eg, age, ethnicity, medication use, depressive symptoms, and anxiety symptoms) were considered in the analysis. The results showed that the effect (decreased marital happiness) increased with increasing levels of sleep disturbance. These findings are consistent with the idea that marital happiness is associated with both the presence and severity (measured by the number of symptoms) of sleep disturbances.

One could have the classic chicken versus egg argument about which comes first being well rested or happily married. Does an unhappy marriage result in reduced sleep or do the negative consequences of sleep deprivation cause marital strife. The research showing a statistical relationship does not indicate which one causes the other. The research simply tells us that marital unhappiness is another of many common results of not enough sleep.

REFERENCES

[1] Landrigan CP, Rothschild JM, Cronin JW, Kaushal R, Burdick E, Katz JT, Lilly CM, Stone PH, Lockley SW, Bates DW, Czeisler CA. Effect of reducing interns' work hours on serious medical errors in intensive care units. N Engl J Med 2004;351:1838–48. for the Harvard Work Hours, Health, and Safety Group.

[2] Fletcher K, Reed D, Arora V. Systematic review of the literature: resident duty hours and related topics. Available from: https://www.acgme.org/acgmeweb/Portals/0/PDFs/Resident_Duty_Hours_and_Related_Topics[1].pdf.

[3] Williamson AM, Feyer A. Moderate sleep deprivation produces impairments in cognitive and motor performance equivalent to legally prescribed levels of alcohol intoxication. Occup Environ Med 2000;57:649–55.

[4] Dawson D, Reid K. Fatigue, alcohol, and performance impairment. Nature 1997;388:235.

[5] Durmer JS, Dinges DF. Neurocognitive consequences of sleep deprivation. Semin Neurol 2005;25(1):117–129. http://faculty.vet.upenn.edu/uep/user_documents/dfd3.pdf.

[6] Short Michelle, Louca Mia. Sleep deprivation leads to mood deficits in healthy adolescents. Sleep Med 2015;16(8):987–93.

[7] Dongen Doran, Dinges. Sustained attention performance during sleep deprivation: evidence of state instability. Archives Italiennes de Biologic 2001;139:253–67.

[8] Zohar D, Tzischinsky O, Epstein R, Lavie P. The effects of sleep loss on medical residents' emotional reactions to work events: a cognitive-energy model. Sleep 2005;28(1).

[9] Kahn-Greene ET, Lipizzi EL, Conrad AK, Kamimori GH, Killgore WDS. Sleep deprivation adversely affects interpersonal responses to frustration. Pers Individ Dif 2006;41:1433–43.

[10] Troxel WM, Buysse DJ, Hall M, Matthews KA. Marital happiness and sleep disturbances in a multi-ethnic sample of middle-aged women. Behav Sleep Med 2009;7(1):2–19.

[11] Troxel WM, Robles TF, Hall M, Buysse DJ. Marital quality and the marital bed: examining the covariation between relationship quality and sleep. Sleep Med Rev 2007;11(5):389–404.

[12] Killgore WDS, Kahn-Greene ET, Lipizzi EL, Newman RA, Kamimori GH, Balkin TJ. Sleep deprivation reduces perceived emotional intelligence and constructive thinking skills. Sleep Med 2008;9:517–26.

[13] Deliens G, Stercq F, Mary A, Slama H, Cleeremans A, et al. Impact of acute sleep deprivation on sarcasm detection. PLoS ONE 2015;10(11):e0140527.

[14] Lim Julian, Dinges, David F. A meta-analysis of the impact of short-term sleep deprivation on cognitive variables. Psychol Bull 2010;136(3):375–89. https://www.med.upenn.edu/uep/user_documents/LimDinges2010MetaAnalysis.pdf.

[15] Van Den Berg J, Neely G. Perceptual and Motor Skills 2006 Performance on a simple reaction time task while sleep deprived. Percept Mot Skills 2006;102:589–99. http://www.amsciepub.com.libproxy.mst.edu/doi/pdf/10.2466/pms.102.2.589-599.

[16] Grissinger M. 'Inattentional blindness' what captures your attention? PT 2012; 37(10):542–555. http://www.ncbi.nlm.nih.gov/pmc/articles/PMC3474444/.

[17] Killgore WD, Balkin TJ, Wesensten NJ. Impaired decision making following 49 h of sleep deprivation. J Sleep Res 2006;15(1):7–13. http://onlinelibrary.wiley.com/doi/10.1111/j.1365-2869.2006.00487.x/full.

[18] Lim J, Dinges D. Sleep deprivation and vigilant attention. Ann New York Acad Sci 2008;1129:305–322. http://onlinelibrary.wiley.com/doi/10.1196/annals.1417.002/epdf.

[19] Donlea JM, Thimgan MS, Suzuki Y, Gottschalk L, Shaw PJ. Inducing sleep by remote control facilitates memory consolidation in Drosophila. Science 2011;332(6037):1571–6.

[20] Mednick Sara, Nakayama Ken, Stickgold Robert. Sleep-dependent learning: a nap is as good as a night. Nat Neurosci 2003;6(7):697–8. http://saramednick.com/htmls/pdfs/Mednick-NN03%5B8%5D.pdf.

[21] Tucker MA, Hirota Y, Wamsley EJ, Lau H, Chaklader A, Fishbein W. A daytime nap containing solely non-REM sleep enhances declarative but not procedural memory. Neurobiol Learn Mem 2006;86(2):241–7.

Fatigue and accidents

Perhaps the most notorious accident associated with human fatigue is the sinking of the MS Herald of Free Enterprise in 1987. The freight ferry sailed from the harbor of Zeebrugge with 459 passengers, 80 crewmembers, and over 100 vehicles on board. The ship capsized minutes after leaving port killing at least 150 passengers and 38 crew members. The Herald capsized because she went to sea with her inner and outer bow doors open. The assistant bosun, Mark Stanley, was responsible to close these doors at the time of departure. He had opened the doors when the ship arrived in port. While the ship was docked he performed normal cleaning and maintenance tasks; after which, he was released from duty and went to his cabin to sleep. He did not hear the call to stations announcement as the ship was being prepared to sail. Others on the crew were either unaware the doors were open or assumed Mr Stanley would report to duty and close them. In fact he remained asleep until he was thrown from his bunk as the ship capsized [1].

This accident was caused by an employee literally asleep on the job but this is not the only type of accident where sleep deprivation and human fatigue contribute. As we have discussed in other chapters, human performance suffers when individuals do not get sufficient, quality sleep. In the following sections we will describe accidents where human fatigue played a role and demonstrate how the negative effects of too little sleep can have deadly consequences.

7.1 BHOPAL—FATIGUE AND POOR ABNORMAL SITUATION RESPONSE

The unintended release of poisonous chemicals from the Union Carbide plant in Bhopal, India on December 3, 1984 remains the world's worst industrial disaster. Estimates of the human impact vary, but the number killed that night is in the thousands. The health of up to half a million more people was severely affected. A debate continues as to the causes and contributing factors of the tragedy. It is clear that human action, whether intentional or not, was a major cause. The response of management contributed to the severity of the incident. There was no clear proof at the time that human fatigue or other factors associated with the night shift operations contributed to the events of that horrible night. In the over 30 years since the explosion, our knowledge of sleep and the effects of fatigue has expanded greatly which has caused many to suspect that it was a contributing factor. There is still debate about what actually happened that night.

Some facts from that fateful night are generally accepted. An estimated 41 tons of deadly methyl-isocynate (MIC) gas leaked out of tank 610 of the Union Carbide plant and escaped into the atmosphere [2]. The MIC production unit was shut down 6 weeks prior to the incident during a plant wide maintenance overhaul. Storage tank 610, containing MIC, had been in isolation at that point [3]. Pressure in the storage tank rose during the third shift on Sunday night Dec. 2–3, 1984. The pressure caused a safety valve to rupture and released the gas into the atmosphere and the surrounding densely populated slum. The pressure building up in the tank was due to an exothermic reaction caused by water in the tank. There is also widespread agreement that the maintenance at the plant was not up to industry standards and was generally neglected. Many have criticized management's indifference to safety and lack of emergency preparedness (Fig. 7.1).

There are two main theories concerning how the water entered the tank. First is the water washing theory; it suggests that a worker was using water to clean lines in another part of the plant and water unintentionally flowed into the storage tank through a series of open valves [4]. Union Carbide's

■ FIGURE 7.1 Photo of accident site in Bhopal, India.

explanation of the accident was deliberate sabotage and someone directly connected a water line to the MIC storage tank [3].

In a journal paper that models the Bhopal tragedy, Eckerman [4] describes how the accident was triggered using the water washing theory as follows:

> *In the evening of 2nd December, at 8:30 p.m., some workers were told to clean the pipes with water. There was no instruction to put in slip binds, so that water could not pass into connecting lines. Because of grossly inadequate maintenance, valves would be malfunctioning and thus permit water to pass even through closed valves. As tank 610 was not holding the pressure, it was possible for large amounts of water to enter the tank this way. The vent lines were made of carbon steel and handling corrosive substances, which would contaminate the tank 610, and catalyze the reaction. When water and contaminants reached the 43 tons of MIC, an exothermic reaction started which caused catalytic trimerisation (a runaway reaction). The temperature and the pressure increased and steam escaped in at least two places. The leakage continued for three hours, until the tank was empty at about two o'clock in the morning.*

At the time of Bhopal disaster, Jackson B. Browning [5] was responsible for Union Carbide Corporation's health, safety, and environmental programs. He directed the teams that responded to and investigated the tragedy and served as a company spokesperson. In a published report, he states the accident had a different cause, sabotage.

> *Although it was not known at the time, the gas was formed when a disgruntled plant employee, apparently bent on spoiling a batch of methyl isocynate, added water to a storage tank. The water caused a reaction that built up heat and pressure in the tank, quickly transforming the chemical compound into a lethal gas that escaped into the cool night air… In March 1985, after three months of work, our technical team told the world that a substantial amount of water had entered the tank, that the water-washing hypothesis was improbable, and that we believed water had entered the tank directly. It took us almost two more years before we could corroborate our scientific findings with interviews and documents.*

The Union Carbide Corporation (UCC) description of what happened was based on an investigative report they sponsored. Ashok Kalelkar of Arthur D. Little, Ltd [3] published a summary of that investigation. It concludes that the water washing of a relief valve vent header causing the disaster is physically impossible. It states that direct water entry into the MIC storage tank was the likely initiating cause of the Bhopal disaster. However, no worker was ever named or charged in this alleged sabotage. It is also reasonable to assume any worker at the chemical plant should have realized how very danger adding water to the MIC tank would be. Regardless of the source of water in the storage tank, management was responsible for overseeing the safety at the plant.

It appears neither the workers nor management at the time understand how serious it was that water had entered the MIC storage tank. At 10:20 pm, shortly before the end of the second shift, the pressure in tank 610 was reported to be at 2 psig [3]. The first sign of the leak was noted between 11:00 and 11:30 pm downwind of the MIC production unit. The MIC supervisor said that he would investigate the leak after tea. The plant supervisors and plant superintendent were taking a break together in the plant's canteen, against company policy, when initial reports of a release were made [3]. About an hour later, the supervisors considered the leakage as "normal" but ordered a water spray onto the leak area. The water washing of the lines was stopped at this time. Around this time the control room operator noted a sudden rise in pressure in the storage tank. A safety valve on the tank opened and by 12:50 am the alarms began inside the factory. Some witnesses claimed the alarms started shortly after tea time began. Kalelkar [3] reports the fire squad sprayed the stack to knock down the gas and reaction subsided in about an hour. Eckerman [4] reports the loud warning siren was stopped at 1:00 am and was not restarted until 2:00 am, about the time most the contents of tank 610 had escaped [6].

Although there is disagreement on the primary cause of the accident (water entering tank 610 due to sabotage vs an error during maintenance), it is clear that human errors were contributing factors. Poor judgment was exercised repeatedly during the hours leading up to the explosion:

■ Reports indicate that the water flushing operations were stopped when the workers observed a blockage. However, 15 min later management gave the order to resume even though the cause of the blockage had not been determined [6].

■ The onsite management seemed more concerned about tea time than the operation of the chemical plant. When initial reports of a leak were received, managers went to the cantina for a break rather than investigating the problem [3].

■ The pressure indicator/control (PIC) showed tank pressure at 10 psig (five times the normal value) around midnight, yet operators ignored these warnings [6].

- There were three safety indictors in the control room that could have signaled operators of problems on the MIC unit. These had been faulty for years [6]. The organization's lack of a safety culture allowed these to be neglected.

Due to financial pressure many UCIL employees had been laid off in 1983–1984. Management curtailed expenditures on maintenance and safety. Staffing for the production operations had been cut from 12 operators to 6 and managers were cut from 2 to 1 [7]. This resulted in a significant increase in workload and a lack of job security among the workers at the Bhopal plant. This combined with shift work and likely fatigued workers probably contributed to the poor judgment displayed as the incident unfolded. Scientists have shown repeatedly that fatigue caused by insufficient sleep and/or shift work has detrimental effects on human performance across multiple tasks, including the ability to process stimuli, make decisions, and increased risk taking. This accident demonstrates with tragic consequences, how a decline in cognitive function due to human fatigue can have devastating results when an abnormal situation arises and humans need to respond.

7.2   AMERICAN AIRLINES 1420—FATIGUE AND DECLINE IN SITUATION AWARENESS

American Airlines flight 1420 crashed after it overran the end of runway during landing at Little Rock National Airport in Little Rock, Arkansas on Jun. 1, 1999 (Fig. 7.2). The airplane struck part of the instrument landing system localizer array (located 411 feet beyond the end of the runway) and then passed through a chain link security fence. The plane continued over a rock embankment and collided with an approach lighting structure and caught fire. The captain and 10 passengers were killed; over 100 passengers were injured. The National Transportation Safety Board (NTSB) determined that contributing to the accident were the flight crew's impaired performance resulting from fatigue and the situational stress associated with the intent to land with poor weather conditions.

The NTSB investigation [8] found that the flight crew's degraded performance was consistent with known effects of fatigue. In this accident they concluded that fatigue had deteriorated the crew's performance on routine landing tasks during the flight's final approach. The crew did not deploy spoilers to slow the plane during landing. The NTSB labeled this error as the single most important factor in the flight crew's inability to stop the plane within the available length of runway. Both crewmembers made basic errors in flight management and the completion of routine tasks, likely due to fatigue. During the final approach the first officer erroneously read a wind report from the air

■ **FIGURE 7.2 Photo of American Airlines 1420, from NTSB.** *http://www.ntsb.gov/ investigations/AccidentReports/Reports/AAR0102.pdf; http://www.planecrashinfo.com/w19990601.htm*

traffic controller as a value within the airline's procedures when in fact the actual wind levels would not have allowed a landing with the crosswinds at the time. In a rush to land, most items on the Before Landing Checklist were not performed. As the crew started the landing process, they did not heed a second warning related to the excessive crosswinds. In the NSTB's judgment during the landing there were cues, such as heavy rain that would have prompted some flight crews to abort the approach. These are examples of cognitive tunneling that can occur with fatigue. The flight cockpit recording documents a conversation between the first officer and the pilot that the landing flaps had not been configured for landing as the pilot thought and that the callouts associated with this task had been skipped when they did not follow the entire Before Landing Checklist, both significant human errors.

In its summary of the flight crew's performance during the landing approach, the NTSB concludes that a collection of changes and weather events should have heightened the crew's awareness of the hazardous conditions. The NTSB [8] report states, "it would have been appropriate for the flight crew to have discussed specific options (holding, diverting to one of the two alternate airports, or performing a missed approach after the airplane was established on the final approach segment) in the event that the weather would necessitate aborting the approach later." The crew had poor situation awareness and decision making which have been shown to be results of sleep deprivation and human fatigue.

7.3 NASA SPACE SHUTTLE—FATIGUE AND DECISION MAKING

Many remember the striking image of Space Shuttle Challenger exploding moments after liftoff in Jan. 1986; killing all seven crew members aboard (Fig. 7.3). The explosion was caused by a hardware failure. The O-ring on the solid rocket booster failed causing the devastating loss for NASA and the country. The primary cause of the accident was faulty decision making and management failures. Contributing to that was sleep deprivation among key decision makers. The O-rings on the external tank had not been tested below 53°. The day of the launch had below-freezing temperatures and engineers recommended the launch be postponed due to the cold and questions related to the effectiveness of the O-rings in such temperatures. The request was rejected by NASA managers, who had slept only 2 h before arriving at work that morning at 1:00 am. The Presidential Commission on the Space Shuttle Challenger Accident noted that time pressure related to the Space Shuttle program increased the potential for sleep loss and judgment errors. The report went on to say that working excessive hours may be admirable, it raises serious questions when critical management decisions are at stake [9].

■ **FIGURE 7.3   NASA Space Shuttle—fatigue and decision making.** *https://en.wikipedia. org/wiki/Space_Shuttle_Challenger*

This was not the first time NASA had such a problem. A potentially catastrophic error occurred just prior to the Challenger explosion when on Jan. 6, 1986, 18,000 pounds of liquid oxygen were inadvertently drained from the Shuttle's External Tank minutes before the scheduled launch. An investigation of the event cited operator fatigue as one of the major contributing factors of the incident. The operators involved had been on duty at the console for 11 h during the third day of working 12-h night shifts. If the launch had not been held the mission would have failed.

> *More important that the average overtime rates was the overtime for certain employees with critical skills. Records show that there was a frequent pattern at Kennedy of combining weeks of consecutive workdays with multiple strings of 11- or 12-hour days. For example, one Lockheed mechanical technician team leader worked 60, 96.5, 94, and 80.8 hours per week in succession during the four weeks ending January 31, 1986. While shiftwork is commonplace in many industrial settings, few can equal a Shuttle launch's potential for inducing pressure to work beyond reasonable overtime limits.*

> *Research has shown that when overtime becomes excessive, worker efficiency decreases and the potential for human error rises. Noteworthy in this regard is Lockheed's review of 264 incidents that caused property damage in 1984 and 1985. More than 50 percent of these incidents were attributable to human error, including procedural deviation, miscommunications and safety violations. On one occasion a potentially catastrophic error occurred just minutes before a scrubbed launch of Shuttle flight 61–C on January 6, 1986, when 18,000 pounds of liquid oxygen were in-advertently drained from the Shuttle's External Tank. The investigation which followed cited operator fatigue as one of the major factors contributing to this incident. The operators had been on duty at the console for eleven hours during the third day of working 12-hour night shift. If the launch had not been held 31 seconds before lift off, the mission might not have achieved orbit [10].*

7.4 EXXON VALDEZ—FATIGUE AND WORK SCHEDULES

On Mar. 24, 1989, the tanker Exxon Valdez grounded in Prince William Sound and spilled about 258,000 barrels of oil causing catastrophic damage to the Alaskan environment. Cost of the cleanup was about $1.85 billion. At the time of the grounding, the third mate was controlling the tanker. The NTSB's Marine Accident Report [11] found that the tanker did not

■**FIGURE 7.4 Photo of the Exxon-Valdez Accident Clean-up.** *Source: https://commons. wikimedia.org/wiki/File:OilCleanupAfterValdezSpill.jpg. This image is in the public domain because it contains materials that originally came from the U.S. National Oceanic and Atmospheric Administration, taken or made as part of an employee's official duties.*

have a sufficient crew which resulted in long work hours and provided little opportunity for sleep for those on the crew. The Safety Board concluded that the third mate could have had as little as 4 h sleep before the beginning of the workday before the accident and only a 1–2 h nap that afternoon. At the time of the accident, he likely had only 5–6 h of sleep in the prior 24 h and was working beyond his normal watch period. The navigation of that portion of the journey was considered high-risk with danger from glacial ice and running aground. At the same time the navigation task was also a very difficult mental task requiring him to recall information about the perimeter of the reef, process information from the radar system to navigate, judging the tanker's position, and decide the timing the tanker's turn. The report [11] states, "He simply may not have been able to incorporate simultaneously all the necessary information for the decision." As the ship started its slow turn, the third mate had by this time realized his error. He did realize his error too late and attempted to make a course correction. The report calls for an aggressive federal program to address fatigue risk in transportation.

Other factors have been cited including a lack of state-of-the-art iceberg monitoring equipment and the radar that was on the ship had been broken for more than a year. The press at the time widely reported that the ship's captain, Joseph Hazelwood, had been drinking heavily the night of the accident. As captain he was in command of the ship while sleeping in his cabin, but the third mate on duty was in control and responsible at the time of the accident. While these factors certainty could have contributed to the accident, the errors made by the third mate have a clear link to sleep deprivation caused by excessive long work shifts and insufficient time for sleep (Fig. 7.4).

7.5 THREE MILE ISLAND AND COGNITIVE TUNNELING

The Three Mile Island (TMI) nuclear reactor near Middletown, Pennsylvania, suffered a partial melt down on Mar. 28, 1979. Even though the amount of radioactive materials released was small and had no detectable public health effect, it is considered the most serious accident of its kind. Various US government reports and books have questioned the human factors, training, and power plant operations. The accident began at around 4 am. Operator fatigue did not cause the accident but likely contributed. Chris Wickens [11] summarizes the accident as being caused by the operators' excessive tunneling on one (incorrect) hypothesis as to the nature of the obvious failure. This tunneling of their attention led them to fail to consider contraindicating information that would have indicated the actual problem and the associated solutions. Cognitive tunneling is common with sleep-deprived individuals.

The accident began when a mechanical or electrical failure stopped the main feedwater pumps from sending water to the steam generators to remove heat from the reactor core. This triggered an automatic shutdown of the turbine generator. Pressure in the primary system began increasing. To control this increase the pilot-operated relief-valve opened. Once the pressure returned to an appropriate level the valve should have closed but it became stuck open. The control room instruments indicated that the valve was closed preventing the operators from realizing cooling water was leaving the system via the open valve. At this point, an excessive number of alarms and warning lightings began. The operators did not realize the problem was a loss of coolant.

Left alone the reactor system would automatically take steps to improve the situation; however, the operators with an incorrect idea of the problem began to take steps to make the situation much worse. They reduced the emergency cooling water that had automatically started flowing into the reactor core which allowed the core to overheat. The accidents at Chernobyl and Fukushima also involved a meltdown of the reactor core. Fortunately, at TMI the containment building remained intact and prevented radioactive release like those at Chernobyl and Fukushima [12].

■ **FIGURE 7.5**   (a) The control room and (b) the towers of the Three Mile Island nuclear facility.

A series of human errors began from the start of the emergency. As the problems began, a control room operator noticed the emergency feed pumps were running, an automated response to improve the situation. What the operator did NOT notice was two lights on the control panel that would have let him know that a different valve was closed on each of the two emergency feedwater lines and the emergency water was not reaching the steam generators. One of the display lights was covered with a yellow maintenance tag. It is not known why the operator did not notice the other warning light [13]. Within minutes there were more than 100 alarms sounding in the control room. An hour into the accident the reactor coolant pumps began to severely vibrate; another indicator of the true problem that went unrecognized by the operators. About 2 h into the accident a conference call was made to the company's vice president and others. With a fresh perspective the people on the phone questioned the status of the stuck valve and identified the problem. Some reports credit a shift supervisor from the day shift that had just arrived at identifying the problem at approximately the same time [13]. Whomever identified the problem they were able to solve it in minutes when the night-shift crew had been struggling with it for 2 h. The people new to the problem were likely well rested and beginning their day unlike the third shift workers (Fig. 7.5).

7.6 METRO-NORTH TRAIN DERAILMENT FATIGUE CAUSED BY CIRCADIAN RHYTHMS AND SLEEP APNEA

A Metro-North train engineer, William Rockefeller Jr., was at the controls of a passenger train when it wrecked on a curve. The accident on Sunday, Dec. 1, 2013, killed four and injured 75. A newspaper at the time reported that Rockefeller stated that he was in "a daze." He said, "I don't know what I was thinking about, and the next thing I know I was hitting the brakes." Prior to applying the brakes the train was going into the curve at nearly three times the speed limit. Rockefeller had worked the afternoon shift for many years and switched to the 5 am shift 2 weeks before the accident [14] (Fig. 7.6).

The government investigation of the accident concluded:

> The National Transportation Safety Board determines that the
> probable cause of the accident was the engineer's noncompliance
> with the 30-mph speed restriction because he had fallen asleep due to
> undiagnosed severe obstructive sleep apnea exacerbated by a recent
> circadian rhythm shift required by his work schedule. Contributing
> to the accident was the absence of a Metro-North Railroad policy
> or a Federal Railroad Administration regulation requiring medical
> screening for sleep disorders. Also contributing to the accident
> was the absence of a positive train control system that would have
> automatically applied the brakes to enforce the speed restriction. [15]

The engineer's personal medical records indicated that he had complained of fatigue prior to the accident. He had several of the risk factors for

■ **FIGURE 7.6 Photo from NTSB investigation.** National Transportation Safety Bureau. http://www.ntsb.gov/investigations/AccidentReports/Reports/RAB1412.pdf

obstructive sleep apnea (OSA) including obesity, snoring, fatigue, and excessive daytime sleepiness [15]. A sleep study after the accident confirmed he had OSA. The engineer was likely asleep or experiencing a microsleep caused by his OSA at the time of the derailment. Changing from the early afternoon shift to an early morning shift required shifting his schedule by 12 h. Such a change can be difficult due to the body's circadian clock. The Thanksgiving holiday days before the accident added to the variations in his sleep schedule.

7.7 FATIGUE'S ROLE IN ACCIDENTS

Most industrial accidents are not caused directly by fatigue or sleep deprivation, however in a shocking number of them, it was a significant contributing factor. These accidents are not due to a single mistake or unusual events. More often it is the combination of poor procedures, atypical events, and faulty human performance. When the out of the ordinary happens a well-rested employee is able to safely address the situation. When the equipment and procedures are well designed and working properly, abnormal events can be managed safely. But when the system is not designed or performing at its best, as was the case in Bhopal and Three Mile Island, the individuals called upon to respond to the situation need to be physically able to call upon their finest judgment and decision-making skills. Sleep-deprived, tired workers will never be at their best. When outside factors such as severe weather in the American Airlines 1420 crash occur, workers' top performance is needed. When events, such as those discussed in the Space Shuttle Challenge Explosion or Exxon Valdez Accident occur, decision makers need to be at their peak. Given that it is impossible to know when these situations may occur it is vitally important to have everyone working at their best day after day. An effective FRMS is key to having well rested, effective workers prepared to handle these situations when they arise.

REFERENCES

[1] Department of Transport, Report of Court No. 8074. mv Herald of Free Enterprise; 1987. https://assets.digital.cabinet-office.gov.uk/media/54c1704ce5274a15b6000025/FormalInvestigation_HeraldofFreeEnterprise-MSA1894.pdf

[2] Bisarya RK, Puri S. The Bhopal Gas Tragedy—a perspective. J Loss Prevent Proc 2005;18(4–6):209–12.

[3] Kalelkar AS, Little AD. Investigation of large-magnitude incidents: Bhopal as a case study; 1988. Available from: http://www.bhopal.com/~/media/Files/Bhopal/casestdy.pdf

[4] Eckerman I. The Bhopal SAGA—causes and consequences of the world's largest industrial disaster. Hydarabad, India: Universities Press; 2005.

[5] Browning JB, 1993. Union Carbide: disaster at Bhopal. Gottschalk JA, editor. Available from: www.environmentportal.in: http://www.environmentportal.in/files/report-1.pdf

[6] Morehouse W, Subramaniam MA. The Bhopal Tragedy: what really happened and what it means for American workers and communities at risk. A Preliminary Report for the Citizens Commission on Bhopal; 1986.

[7] Shrivastava P, Mitroff D, Miglani A. Understanding industrial crises. J Manage Stud 1988;25:4.

[8] National Transportation Safety Board Aircraft Accident Report, Runway Overrun During Landing American Airlines Flight 1420 Little Rock, Arkansas June 1, 1999. PB2001-910402 NTSB/ARR-01/02 DCA99MA060. Available from: http://www.ntsb.gov/investigations/AccidentReports/Reports/AAR0102.pdf

[9] http://www.rawstory.com/2015/10/the-worlds-worst-disasters-caused-by-sleep-deprivation/

[10] https://www.gpo.gov/fdsys/pkg/GPO-CRPT-99hrpt1016/pdf/GPO-CRPT-99hrpt1016.pdf

[11] http://www.aviation.illinois.edu/avimain/papers/research/pub_pdfs/techreports/05-23.pdf

[12] http://www.nrc.gov/reading-rm/doc-collections/fact-sheets/3mile-isle.html.

[13] Report of The President's Commission on The Accident at Three Mile Island, October 1979, Washington DC. Available from: http://www.threemileisland.org/downloads/188.pdf

[14] http://www.nydailynews.com/new-york/metro-north-train-engineer-daydreaming-crash-article-1.1535870?utm_content=buffer8b022&utm_source=buffer&utm_medium=twitter&utm_campaign=Buffer#ixzz2mS3UIrrd

[15] NTSB report. Available from: http://www.ntsb.gov/investigations/accidentreports/pages/RAB1412.aspx

Fatigue-related regulations and guidelines

8.1 OSHA AND FATIGUE RISK

The Occupational Safety and Health Administration (OSHA) is an agency of the US Department of Labor. It was founded in 1970 by an act of Congress. OSHA's mission is to assure safe working conditions in the United States by enforcing numerous safety regulations. There are some exceptions such as those for the self-employed or family farms, but most private sector employers are bound by OSHA regulations. OSHA has jurisdiction unless preempted by another federal agency such as the Department of Transportation or Federal Aviation Administration, but these agencies can only preempt OSHA in a specified activity or task. OSHA has the ultimate responsibility for the safety and health of American employees.

OSHA does not have standards currently that address fatigue risk management. It has taken a position on extended work shifts which includes the definition of an extended shift as:

> *A normal work shift is generally considered to be a work period of no more than eight consecutive hours during the day, five days a week with at least an eight-hour rest. Any shift that incorporates more continuous hours, requires more consecutive days of work, or requires work during the evening should be considered extended or unusual. Extended shifts may be used to maximize scarce resources. Long or unusual shifts are often required during response and recovery phases of emergency situations such as terrorist threats, which generally come without warning, require continuous monitoring, and may overwhelm local responders both technically and tactically. These schedules ensure that the appropriate scarce resources are in place and accessible while full mobilization is being developed.*
>
> **OSHA website [1]**

Statement

U.S. Department of Labor

Release Number: 10-1238-NAT
Sept. 2, 2010
Contact: Jason Surbey Diana Petterson
Phone: 202-693-4668 202-693-1898
E-mail: surbey.jason@dol.gov petterson.diana@dol.gov

Statement by US Department of Labor's OSHA Assistant Secretary
Dr. David Michaels on long work hours, fatigue and worker safety

WASHINGTON - The U.S. Department of Labor's Occupational Safety and Health Administration has been petitioned by Public Citizen, a national advocacy organization, as well as other groups and individuals, to issue regulations that would limit the work hours of resident physicians. In response to the request, the assistant secretary of labor for occupational safety and health, Dr. David Michaels, today issued the following statement:

"We are very concerned about medical residents working extremely long hours, and we know of evidence linking sleep deprivation with an increased risk of needle sticks, puncture wounds, lacerations, medical errors and motor vehicle accidents. We will review and consider the petition on this subject submitted by Public Citizen and others.

"The relationship of long hours, worker fatigue and safety is a concern beyond medical residents, since there is extensive evidence linking fatigue with operator error. In its investigation of the root causes of the BP Texas City oil refinery explosion in 2005, in which 15 workers were killed and approximately 170 injured, the U.S. Chemical Safety Board identified worker fatigue and long work hours as a likely contributing factor to the explosion.

"It is clear that long work hours can lead to tragic mistakes, endangering workers, patients and the public. All employers must recognize and prevent workplace hazards. That is the law. Hospitals and medical training programs are not exempt from ensuring that their employees' health and safety are protected.

"OSHA is working every day to ensure that employers provide not just jobs, but good, safe jobs. No worker, whether low-skilled and low-wage, or highly trained, should be injured, or lose his or her life for a paycheck."

Under the Occupational Safety and Health Act of 1970, employers are responsible for providing safe and healthful workplaces for their employees. OSHA's role is to assure these conditions for America's working men and women by setting and enforcing standards, and providing training, education and assistance. For more information, visit http://www.osha.gov.

#

■ **FIGURE 8.1 OSHA's statement on hours of service.** *(Source: https://www.osha.gov/pls/oshaweb/owadisp.show_document?p_table=NEWS_RELEASES&p_id=18285).*

OSHA does not have specific regulations about extended shift work. The agency's website does provide information on fatigue risk management as a guide. There is potential that OSHA could adopt standards in this area. They issued a statement in 2010 (see Fig. 8.1) supporting the report of US Chemical Safety and Hazard Investigation Board (CSB) on the BP Texas City accident. The statement clearly shows OSHA's belief that long work hours can be hazardous. Over the years the agency has written citations of safety infractions that are not specifically covered by regulations using the general duty clause which states "Each employer shall furnish to each of his employees employment and a place of employment which are free from recognized hazards that are causing or are likely to cause death or serious physical harm to his employees." It is plausible that in the future when OSHA is investigating a serious accident that excessive work hours could be cited as causing an unsafe workplace and the agency could levy fines for this.

8.2 NIOSH SLEEP-RELATED PUBLICATIONS

At the same time as OSHA was established by the US Congress to regulate workplace safety another government organization was founded: the National Institute for Occupational Safety and Health (NIOSH). Its mission is to develop and transfer knowledge in the field of occupational safety and health. It does not have OSHA's role of writing regulations, but it has published

Fatigue Prevention

Disaster response workers often work longer shifts and more consecutive shifts than the typical 40-hour work week. Working longer hours may increase the risk of work injuries and accidents and can contribute to poor health. The scientific literature indicates that working at least 12 hours per day was associated with a 37% increased risk of injury and construction workers working more than 8 hours per day had a 15 % higher injury rate. Fatigue and stress from strenuous work schedules can be compounded by heavy physical workloads, unfavorable environmental conditions (e.g., damaged infrastructure, hazardous materials and debris, sparse living conditions), long commutes, and personal demands on workers.

Therefore, disaster response organizations should have management plans in place to minimize fatigue risks, recognize hazards, and provide regular opportunities for worker rest and recovery. The National Response Team, an organization of 15 Federal departments and agencies responsible for coordinating emergency preparedness and disaster response, has prepared guidance and checklists for managing fatigue during disaster recovery operations.

Consider the following general guidelines:
- ***Regular Rest:*** Establish at least 10 consecutive hours per day of protected time off-duty in order to obtain 7-8 hours of sleep. Rest and a full complement of daily recovery sleep are the best protections against excessive fatigue in sustained operations. Allowing only shorter off-duty periods (e.g. 4-5 hours) can compound the fatigue of long work hours.
- ***Rest Breaks:*** Frequent brief rest breaks (e.g., every 1-2 hours) during demanding work are more effective against fatigue than few longer breaks. Allow longer breaks for meals.
- ***Shift Lengths:*** Five 8-hour shifts or four 10-hour shifts per week are usually tolerable. Depending on the workload, twelve-hour days may be tolerable with more frequent interspersed rest days. Shorter shifts (e.g. 8 hours), during the evening and night, are better tolerated than longer shifts. Fatigue is intensified by night work because of nighttime drowsiness and inadequate daytime sleep.
- ***Workload:*** Examine work demands with respect to shift length. Twelve-hour shifts are more tolerable for "lighter" tasks (e.g., desk work). Shorter work shifts help counteract fatigue from highly cognitive or emotionally intense work, physical exertion, extreme environments, or exposure to other health or safety hazards.
- ***Rest Days:*** Plan one or two full days of rest to follow five consecutive 8-hour shifts or four 10-hour shifts. Consider two rest days after three consecutive 12-hour shifts.

■ **FIGURE 8.2 NIOSH's fatigue prevention guidance for the deepwater horizon response.** *(Source: http://www.cdc.gov/niosh/topics/ oilspillresponse/protecting/default.html#effects).*

guidelines about shift work. The publication *Plain Language about Shift-work* [2] is available free of charge on the NIOSH website. In the publication they acknowledge that overly tired workers are more likely to make errors and cause accidents presenting a safety hazard. Reducing the hazards associated with shift work is discussed in detail in chapter: Work Shifts.

In 2010, NIOSH released *Interim Guidance for Protecting Deepwater Horizon Response Workers and Volunteers* [3]. In this publication, they addressed fatigue prevention and made recommendations related to rest breaks, shift lengths, workload, and days off. Fig. 8.2 is the section on fatigue prevention from that document. The document shows that NIOSH believes longer work shifts results in increased risks. It also states NIOSH's recommendations for fatigue prevention. It is reasonable to expect NIOSH to continue developing guidelines related to fatigue risk management and that future NIOSH documents will extend recommendations beyond this particular disaster response. In fact, a 2012 NIOSH science blog highlighted several ongoing projects to reduce the risks associated with long working hours and shift work. This includes exploring new methods to better measure work hours and to quantitatively estimate the associated risks. This research is to support policies and recommendations targeted at reducing illness and injuries associated with long work hours, shift work, and irregular work schedules. It is likely that OSHA will follow with associated regulations in this area.

Another area of interest for NIOSH is effective training interventions related to fatigue risk. In 2014 and 2015 [4], NIOSH developed training related to sleep deprivation and shift work for two specific industries, nursing and emergency responders. They have announced they are currently developing training in this area for the mining industry. It is likely the institute will continue to develop training material related to fatigue risk and to make it freely available online.

NIOSH TRAINING

NIOSH offers two free online courses, WB2408 and WB2409, to educate nurses and their managers about the about the health and safety risks associated with shift work, long work hours, and related workplace fatigue issues. The training also covers workplace strategies and personal behaviors to reduce these risks. The training includes a post test. Continuing education certificates are available with the training. There is similar interim NIOSH training for emergency responders working long hours also available for free on the NIOSH website.

8.3 UK AND EU REGULATIONS

The Health and Safety Executive (HSE) in the United Kingdom has regulations on work hours. These regulations include:

- Health and Safety at Work Act 1974
- Management of Health and Safety at Work Regulations 1999
- Working Time Regulations 1998

The HSE has also published a guideline on the subject, *Managing Shift Work: Health and Safety Guidance* HSG 256 [5,6], that is available for free online.

The 1998 Working Time Regulations is a detailed document and those obligated to comply with the requirements should refer to the document directly. A brief summary of the rights and protections for workers is:

- a limit of an average of 48 h per week which a worker can be required to work (though workers can choose to work more if they want to)
- for night workers, a limit of an average of 8 h work in each 24-h period
- a right for night workers to receive free health assessments
- a right to 11 h consecutive rest a day
- a right to 1 day off each week
- a right to a rest break if the working day is longer than 6 h
- a right to 4 weeks' paid leave per year

The United Kingdom's HSE also provides information related to shift work offshore in its 2008 information sheet titled *Guidance for Managing Shiftwork and Fatigue Offshore* [7,8]. The Management of Health and Safety at Work Regulations is similar to OSHA's regulations in many ways. Under it, employers are required assess the risks to employees from work activities and make a commitment to make "reasonably practicable" measures to manage these risks. This includes the number of hours worked and how these hours are scheduled. Employees also have a responsibility to take reasonable care of their own health and safety and that of other people, who may be affected by their activities at work. This includes managing fatigue-related risk factors [9].

These UK publications highlight that safety guidelines on fatigue risk management tend to be similar between countries. However, there are specifics in the regulations that vary location to location. The European Commission on Mobility and Transport has worked to establish procedures and requirements for drivers across multiple countries. The European Aviation Safety Agency also has fatigue management in aviation and its newest fatigue management regulations (EU Commission Regulation No. 83/2014) go into effect in 2016.

8.4 TRANSPORTATION FATIGUE REGULATIONS

Transportation was one of the first industries to realize the risks associated with fatigued workers and to regulate hours of work and rest as far back as the 1940s. In the United States, OSHA has jurisdiction over off-highway loading and unloading, such as warehouses, plants, terminals, and retail locations in the United States. The US Department of Transportation (DOT) has jurisdiction over interstate highway driving, commercial driving licensing, hours of service, and roadworthiness of the vehicles. The Federal Aviation Administration (FAA) regulates flight crews and some other aspects of the safety of ground crews. All of these organizations, and their counterparts in other countries, have regulations to reduce the risks associated with human fatigue.

Aviation regulatory agencies have an additional challenge from changes in time zones as flight crews travel across the globe. The FAA domestic flight rules do not limit the hours on duty for pilots. Rather, they regulate flight time limitations and required rest periods. Pilots of domestic flights are generally limited to 8 h of flight time during a 24-h period. This may be extended if the pilot receives additional rest time after completing flying. A pilot cannot fly if he or she has not has at least 8 h of continuous rest during the preceding 24-h period. If a pilot had fewer than 9 h rest in the preceding 24-h period, the next rest period must be longer to compensate. In recent years, the FAA has also required dispatchers, schedulers, and those working in operational control be covered in a fatigue risk management system (FRMS). The FAA requirements can be found in *14CFR 117.7, 121.473, and 121.527 Fatigue risk management system.*

Similar European Union requirements can be found in the Commission Regulation No. 83/2014. This regulation that goes into effect in 2016 and is a significant change in the method used to management the risks associated with human fatigue. Previously, the European Union (EU) regulated hour of service for flight crews. As has been the trend in varying industries, the new EU regulation moves from hours of service to an FRMS approach. It is a performance-based regulation with an oversight approach that focuses on measurable safety outcomes, rather than solely using prescriptive time limits for crew members. The regulation calls for the implementation safety management systems approach and the use of FRMSs.

The Federal Railroad Administration (FRA) via the *National Rail Safety Action Plan Final Report 2005–2008* highlighted the problem of fatigue among railroad operating employees. The FRA called for improved crew scheduling to manage the risk of fatigue in the railroad industry. The FRA acknowledges that an approach that limits hours of service is necessary but not sufficient to prevent operator fatigue. The FRA has called for proactive fatigue risk management programs that balance the amount of work, when it is performed, and other factors much like other transportation regulatory agencies.

The US DOT has multiple regulations related to fatigue risk management. *49CFR 192.631 Control Room Management* applies to control room operators of pipeline facilities. This regulation calls for a written plan that includes a fatigue mitigation plan. A portion of the CFR is shown in Fig. 8.3.

Each operator must implement the following methods to reduce the risk associated with controller fatigue that could inhibit a controller's ability to carry out the roles and responsibilities the operator has defined:

(1) Establish shift lengths and schedule rotations that provide controllers off-duty time sufficient to achieve eight hours of continuous sleep;

(2) Educate controllers and supervisors in fatigue mitigation strategies and how off-duty activities contribute to fatigue;

(3) Train controllers and supervisors to recognize the effects of fatigue; and

(4) Establish a maximum limit on controller hours-of-service, which may provide for an emergency deviation from the maximum limit if necessary for the safe operation of a pipeline facility.

■ **FIGURE 8.3 Pipeline control room fatigue mitigation regulation.** *(Source: From 49CFR 192.631.).*

The Federal Motor Carrier Safety Administration has hours of service regulation for drivers in *49CFR 395.1*. The regulation is very detailed and has differing regulations for various situations. In general the regulation sets a driving limit of 11 h after 10 consecutive hours off duty for property-carrying drivers and 10 h after 8 consecutive hours off duty for passenger-carrying drivers. Cumulative driving time is also regulated. Drivers may not drive after 60 h of duty in 7 consecutive days or 70 h of duty in 8 consecutive days.

DROWSY DRIVERS

New Jersey has a law that a sleep-deprived driver qualifies as a reckless driver who can be convicted of vehicular homicide. It is named Maggie's Law, in honor of college student who was killed in a head-on crash caused by a driver who had not slept in the 30 h prior to the accident. The driver's defense was that there was no law against falling asleep while driving and he had done nothing wrong. The trial judge accepted this argument and the driver received a suspended jail sentence and a $200 fine. In response, the victim's family successfully lobbied for a new law that defines fatigue as being without sleep for more than 24 consecutive hours and makes driving while fatigued a criminal offense.

8.5 HEALTHCARE FATIGUE REGULATIONS

In 2001, [10] a petition was filed by a union, the Committee of Interns and Residents, and Public Citizen, a consumer advocacy group, requesting OSHA to limit the number of hours medical residents work. They stated that residents regularly work 95 h per week and some times as many as 136 h per week. The groups requested limits on hours worked, frequency of night shifts, and a minimum block of off-duty time per week. OSHA did not agree to the request and defer to the Accreditation Council for Graduate Medical Education (ACGME). In 2010, [11] these groups again petition OSHA for regulations in this area. The ACGME published a new standard addressing this issue in 2011 [12]. The standard has the following requirements for hours of service, off-duty time, maximum duty period, night shifts, and time off:

- Duty hours must be limited to 80 h per week, averaged over a 4-week period, inclusive of all in-house call activities and all moonlighting.
- Residents must be scheduled for a minimum of 1 day free of duty every week (when averaged over 4 weeks). Home calls cannot be assigned on these free days.
- Duty periods of first-year residents must not exceed 16 h. More senior residents are limited to 24 h of continuous in-house duty.
- Residents should have 10 h, and must have 8 h, free of duty between scheduled duty periods. Intermediate-level residents must have at least 14 h free of duty after 24 h of in-house duty.
- Residents must not be scheduled for more than six consecutive nights of night float.
- A mandatory 4 days off per month and 1 day off per week averaged over 4 weeks is required.

In general, these regulations have been seen as an improvement for medical residents and healthcare as a whole. Some, however, have criticized that the hours of work are averaged which can still result in a high number of work hours in a single week if other weeks in the time period have less.

8.6 **CONCLUSIONS**

In the field of safety, government regulations are the minimum level of safety that is acceptable. These rules are the level of safety that is specifically defined and enforced by inspections and fines. This level is the lowest possible and reflects an organization that does not have a safety culture and does not value the well-being of their workers. A higher level of safety would be operating at the industry standard. This level is applying common work practices and procedures that are well established but have yet to be codified as regulatory standards. It does reflect an organization that has a safety culture and a respect for humans. Decisions are governed by not only what is required but also what should be done to provide a safe workplace and minimize the risks associated with doing business. The highest level of safety is going beyond what is the industry standard and innovating safety. Such organizations should consider the best practices from other industries and adapting it for their industry. They explore what can be done to improve workplace safety and reduce risks and are leaders in improving their industry.

In the area of fatigue risk management in the processing industry, the lowest level of safety is managing hours of service. This satisfies regulatory requirements in many situations and can potentially avoid fines. The industry standard for the chemical processing industry, however, has moved to FRMSs. It is more than tracking work hours. It is training on the importance of sleep. It is identifying and managing sleep disorders among workers. It is change the work environment and the work task to reduce monotony and increase employee vigilance.

Regulatory agencies in the United States and elsewhere are moving toward FRMS being the required method of addressing the risks associated with human fatigue. Companies with a safety culture of just following the governmental regulations should prepare because FRMS requirements are coming. Companies with a solid safety culture should adopt a FRMS now. It is a well-established management practice that reduces fatigue related risks; potentially saving lives and preventing costly accidents.

REFERENCES

[1] https://www.osha.gov/SLTC/emergencypreparedness/guides/extended.html
[2] http://www.cdc.gov/niosh/docs/2004-143/pdfs/2004-143.pdf Plain language about shiftwork
[3] http://www.cdc.gov/niosh/topics/oilspillresponse/protecting/default.html#effects
[4] NIOSH, Caruso CC, Geiger-Brown J, Takahashi M, Trinkoff A, Nakata A. NIOSH training for nurses on shift work and long work hours. (DHHS (NIOSH) Publication No. 2015-115). Cincinnati, OH: US Department of Health and Human Services, Centers for Disease Control and Prevention, National Institute for Occupational Safety and Health. [www.cdc.gov/niosh/docs/2015-115/]; 2015.
[5] http://www.hse.gov.uk/pubns/priced/hsg256.pdf Managing shift work: health and safety guidance HSG 256.
[6] http://books.hse.gov.uk/hse/public/saleproduct.jsf?catalogueCode=9780717661978
[7] From http://www.hse.gov.uk/offshore/infosheets/is7-2008.htm
[8] http://www.hse.gov.uk/foi/internalops/hid_circs/enforcement/spcenf160.htm
[9] From http://www.hse.gov.uk/humanfactors/resources/articles/api-fatigue-standard.htm
[10] http://ehstoday.com/news/ehs_imp_34385
[11] https://www.aamc.org/advocacy/washhigh/highlights2011/167864/osha_reviews_petition_to_regulate_resident_hours.html
[12] https://www.acgme.org/acgmeweb/Portals/0/PDFs/jgme-monograph[1].pdf

Chapter 9

Fatigue counter measures

Human fatigue can have significant safety consequences. Society is gradually becoming aware of the importance of sleep and the many dangers caused by people who are sleep deprived. The transportation industry has a long history of addressing this risk. Other industries are beginning to realize the importance of human fatigue risk management. There are many steps organizations can and should take to address the hazards associated with fatigue. There are also steps the individual should take. In this chapter we will focus on the counter measures a worker can take to improve their sleep, manage their fatigue, and increase their alertness.

9.1 **SCHEDULE**

The human body is very much driven by an internal clock. Our circadian rhythms guide our body to be more alert at times and better suited for sleep at other times. There are more details concerning how circadian rhythms and sleep debt interact in a prior chapter. One key to countering fatigue is managing one's schedule.

Keep a consistent sleep schedule: Your body's circadian rhythms promote sleep and wake at approximately the same time each day. Periodically pulling an all-nighter wreaks havoc on your internal schedule. Maintaining a schedule that widely varies can result in the situation when your body has to try to sleep when the circadian system is promoting wakefulness. Under these circumstances, either one will be unable to fall asleep or have built up a dangerous amount of sleep debt to be able to sleep. A consistent schedule will help coordinate sleep and wake at the necessary times.

Use anchor sleep: When you cannot follow a consistent schedule due to shift work or varying work demands, try to keep a portion of your sleep at the same time. For example, if your normal sleep schedule is 10 pm to 5 am, try to sleep during a portion of that span when you are working a different shift. If you had to work some evening shifts until midnight, going to bed at 1 am until 8 am will give you a 4-h period of anchor sleep from 1 am until

5 am. The use of anchor sleeping periods will help you manage shift transitions and days off with less impact on your ability to get sleep and maintain cognitive performance.

Extend your day, don't shorten it: A common tip to avoid jetlag is to stay awake and adjust to the local time rather than taking a nap when you first land. This allows you to synch your circadian rhythms, sleep debt, and local time zone. If you need to get up 2 h early, your body will fight you if you try to go to sleep 2 h early to compensate.

Schedule your tasks around your peak time: If you can, do your most boring tasks early in the day when your mind is more alert and save your more interesting tasks for after lunch when your mind and body may start to drag. Performing physical activities when your energy levels are at their lowest can help as well.

9.1.1 Exercise

Doctors advocate regular exercise. Vigorous exercise soon after you wake up raises body temperature and makes you more alert. It can synchronize your body's clock. On the other hand, avoid doing exercise within an hour or two of bedtime. It keeps body temperature evaluated and makes falling asleep more difficult.

9.1.2 Naps

A short nap can do wonders to fight off fatigue. Keeping naps to about 20 min is recommended. Longer naps can result in sleep inertia. Don't nap too close to your normal bedtime; the decrease in sleep debt from the nap can make it harder to fall asleep.

9.2 FOOD AND DRINK

One of the cues our body uses for circadian rhythms is the food we eat. What we eat or drink causes a response from our body including how well we sleep or don't sleep.

Plan when you eat: Try to eat breakfast soon after you get out of bed. This tells your body's clock that the day is starting. Avoid spicy, hard-to-digest foods before you are planning on going to bed. If you must shift your sleep times, adjust your mealtimes to follow the changes.

Use caffeine cautiously: It takes about 20 min for the stimulating effects of caffeine to kick and they can last for 8–10 h. Avoid caffeine in the hours before bedtime to get your best sleep.

Do not drink alcohol to get to sleep: Alcohol makes sleep shorter and more restless. It may make it easier to fall asleep but you won't get a good restful night's sleep under the influence.

Avoid drinking too many liquids at night: Drinking a lot of fluids may cause the need for frequent bathroom trips interrupting sleep throughout the night. Caffeinated drinks are a diuretics which adds to the problem. Use the restroom before going to bed and reduce the need to use the toilet during the night.

Do not smoke at bedtime: Nicotine is a stimulant and will make it harder to sleep.

9.3 A SLEEP-FRIENDLY BEDROOM

Your bedroom can be conducive for sleep or make it harder to fall asleep and stay asleep.

Make it dark: Dark bedrooms with opaque shades or curtains will make sleep easier to come by especially for those trying to sleep during the day. If someone else needs light in the bedroom while you are trying to sleep an eye mask is another option to block out light.

Avoid electronic screens: Exposure to light at night can interfere with your body's circadian rhythms and the ability to get to sleep. Blue light emitted by electronics is especially disruptive. The screens on cell phones, tablets, computers, and TVs are a major source for this type of light. Another concern is if you are watching something stimulating rather than relaxing. Reading a book can be relaxing, but a backlit device is more disruptive than a one illuminated from the front.

Make the place quiet to sleep: Try to minimize noise as much as possible. Silence or move your cell phone away from your bed to avoid noises while you sleep. Earplugs can be useful in some situations. For those trying to sleep during the day, putting a sign on the doorbell can warn others not to disturb you. Another option is using white noise to mask other noise or a cd of sounds from nature playing in the background as you doze off.

Keep your bedroom cool: Room temperature can also affect sleep. Most people sleep best in a slightly cool room, around 65°F or 18°C. A bedroom that is too hot or too cold can interfere with quality sleep.

Wear comfy pajamas: You want sleep clothing that is loose and soft. Avoid buttons and other things that might poke you as you reposition in your sleep. Sleeping with socks can help some individuals feel warm and sleepy.

BED PILLOW ADVICE

The wrong pillow may worsen headaches, neck pain, arm numbness, and allergies symptoms. As we sleep, our body sheds skin cells which can result in pillows collecting dead skin cells, mold and dust mites. A general rule is to buy new bed pillows every 12–18 months.

Ideally your pillow should keep your neck in a neutral alignment, not bent too far forward or backward. Your sleep posture dictate the type of pillow that is best for sleep.

- Back sleepers: Use a thinner pillow and avoid having your head tilted too far forward, with your chin near your chest. A pillow that is thicker at the bottom can cradle your neck while not tilting your head.
- Side sleepers: Use a firmer pillow to fill to support the head uniformly. Avoid the top of your head tilting toward or away from the bed.
- Stomach sleepers: Use a very thin, flat pillow. You may not even need a pillow.

A bonus to sleeping with the right pillow is that you might snore less, which is bound to make anyone else in the bed happier.

Use comfortable beding: You should have enough room to stretch and turn comfortably without becoming tangled in your bed covers. Waking up with a sore back or an aching neck, may be a sign you need a new mattress or pillow.

Keep your pets off the bed: Many of us love our cats and dogs. However, as they move and reposition during the night they may be affecting your sleep by waking you up several times a night, even if you don't realize it.

9.4 LIGHTING

Light has a significant important effect on the human body, it synchronizes our internal body clock to the day. Bright light gives our body the signal it is time to be awake, while our bodies associate dark with sleep. Keeping this in mind can help you use light to wake up or sleep. For example, night workers who plan to go to bed when they get home might want to wear sunglasses to avoid the bright sunlight as they travel home from signaling their body to stay awake. Oppositely, going outside on a sunny day during a work break can help arouse someone from feelings of sleepiness. Light also synchronizes the circadian rhythms. If one is trying to shift their wake up point to an earlier time, a light box may be useful. Bright light for about 30 min at the beginning of the day can help move one's circadian rhythms to a desired time.

9.5   **GETTING TO SLEEP OR BACK TO SLEEP**

It is common to wake briefly multiple times during the night; many of us don't even remember it in the morning. If you have trouble falling asleep or back to sleep there are several tips that might help.

Have a relaxing bedtime routine: Allow yourself time to wind down and relax before going to bed. Following a set routine can help signal your mind it is time to get ready to sleep.

Relax your body and mind: Signal your body you are ready for sleep. Stay in a relaxed posture and avoid stressing or thinking things that make you anxious. Breathing exercises can help "turn off" your mind. Instead of thinking about your hectic day or endless to-do list, focus on your breath. Repeating breath in and out slowly and let your mind slow down. Other relaxation technique including visualization, progressive muscle relaxation, and meditation can help you get to sleep. Others swear by simple word or number games like Sudoku to unwind their mind before sleep.

Aromatherapy: Proponents of essential oils claim the fragrance of lavender can induce a deep sleep. The fragrance can be used with scent oil in a diffuser.

KEEPING A SLEEP JOURNAL

If you are having sleep related problems, keeping a sleep journal can help you identify potential sources and possible solutions. It can also provide useful information for a sleep professional in diagnosing problems. There are online apps for sleep journals, in addition to free charts that can be printed. In general you want to track the following:

    When you went to bed and woke up
    How long and well you slept
    If you were awake during the night
    What and when you ate and drank
    What and when you took medications
    What emotion or stress you had
    What exercise you had during the day
    If you took a nap and how long it was
    Were you tired during your typical daily activities

There are also gadgets and apps that monitor you while you sleep. They claim to measure a variety of things related to sleep including how much you toss and turn, calculate your sleep debt, and listening for snoring. Most of their claims have not been scientifically validated but they might still be worth checking out.

Do a quiet activity: If you have tried for 15 min to get to sleep, try getting out of bed and doing a quiet activity. Reading a book is a common technique for getting to sleep.

Write it down and forget about it: Some experts suggest keeping a notebook beside your bed if you frequently wake up feeling anxious or with your mind racing. Making a brief note about what is on your mind can signal your mind to relax. Knowing you will remember the great idea in the morning can help you relax.

9.6 CONCLUSIONS

This chapter has summarized a variety of steps an individual can take to manage fatigue and achieve better sleep. Previously chapters in the book have more detailed explanations about the biology of sleep. These can be useful to understand how and why some of these tips work. However, it is important to remember that the majority of the burden for fatigue risk management is on the employer rather than the employee. If a work schedule is such that a worker does not have adequate time to sleep none of the suggestions in this chapter will solve the problem or reduce the risks associated with human fatigue. Successful fatigue risk management needs to be a joint effort of everyone involved.

Chapter

10

Work shifts

10.1 SHIFT WORK

Business is no longer just 8–5. Technology and the speed of the modern world have driven our need for workers on multiple shifts. It is often cost prohibitive to have expensive equipment used only 8 h a day. Critical operations often require around the clock operators and supervision. There are numerous work shift and schedule combinations that can be used to meet these needs. Each has different features, advantages, and disadvantages. With the complexity of the modern workplace, no single shift system is best in all situations. However, a well-planned system can improve the health and safety of workers and their satisfaction. Shift timing, duration, rotation, and work breaks are key factors in work schedules. Other aspects of the job should be considered when developing schedules including required staffing levels, hand-offs between shifts, and how work will be divided between work shifts. Considerations outside the organization can influence work schedules. Examples include arranging start and finish times to be convenient for public transportation and family commitments such as school starting times and daycare options. In some situations, there will be requirements on work hours, both minimum and maximum based on contractual labor agreements, corporation policies, and government regulations. Chapter 8, Fatigue-Related Regulations and Guidelines, has a detailed discussion of regulations and guidelines that should be considered when developing a work scheduling system. In the literature on works shifts there are some terms that are commonly used. These include the following:

- Shiftwork: Working outside of normal daylight hours, outside 7 am to 6 pm
- Rotating shifts: Working at different, changing shift times. Shift changes may be weekly or monthly
- Permanent shifts: Working one set shift without change
- Day shift: Typically starts between 5 and 8 am and ends between 2 and 6 pm, also called first shift

Table 10.1 Examples of Work Shift Designs

Shift Designs	Work Days/Off Days	Shift Sequence
Permanent day shift	5/2	D-D-D-D-D-f-f, D-D-D-D-D-f-f, …
Permanent evening shift	5/2	E-E-E-E-E-f-f, E-E-E-E-E-f-f, …
Alternating day/evening shifts with 1 week rotation	10/4	D-D-D-D-D-f-f, E-E-E-E-E-f-f, …
Alternating day/evening, night with 1 week forward rotation	15/6	D-D-D-D-D-f-f, E-E-E-E-E-f-f, N-N-N-N-N-f-f, …
Alternating day/evening, night with 1 week backward rotation	15/6	D-D-D-D-D-f-f, N-N-N-N-N-f-f, E-E-E-E-E-f-f, …
Metropolitan rotation	6/2	D-D-E-E-N-N-f-f, …
Continental rotation	21/7	D-D-E-E-N-N-N, f-f-D-D-E-E-E, N-N-f-f-D-D-D, E-E-N-N-f-f-f, …

D, day shift; E, evening shift; N, night shift; f, free day.
Adapted from Kroemer [1].

- Evening shift: Typically starts between 2 and 6 pm and ends between 10 pm and 2 am, also called second shift
- Night shift: Typically starts 10 pm–2 am and ends 5–8 am, also called the graveyard shift or third shift
- Split shifts: When an individual's daily work is divided into two or more shifts
- Flextime: Is a flexible arrangement where employees work a set amount of time (8 h) during a longer block of time (10 h), all workers overlap on a core amount of time in the midday

The most common system in western economies is the same 8-h shift for 5 consecutive days with the following 2 days free time for the employee [1]. Table 10.1 provides examples of different shift designs. All of the designs provide 2 consecutive free days, although these are not always on Saturday and Sunday.

10.2 WORK-SHIFT SCHEDULE DESIGN

A well-designed work schedule will provide adequate rest time, reduce the risks associated with human fatigue, and support employee morale. Developing effective schedules requires balancing social, personal, psychological, medical, and safety concerns. There are several important considerations for organizations designing work shifts.

Length of the rotation period (the number of days on any one shift before switching to the next shift) is a major consideration in work shift schedule design. Many organizations rotate shifts after a period of 1 week, with 5–7 consecutive night shifts. However, since it generally takes at least 7 days for adjustment of the circadian rhythms, this system "fights" the body's circadian rhythms. Just as the body adjustment to a new shift starts, it is time to rotate to the next shift that is again out of synch with the body's nature rhythms [2]. Longer shift rotation periods such as 2 weeks or 1 month on the same shift would allow the body's circadian rhythms more time to adjust. Challenges for workers on rotating shifts can be the desire to return to "normal" day/night schedule on days off potentially cancelling any adaptation from the body. Longer rotation periods can also result in longer periods of social isolation may cause more disruption in employee's personal lives.

In some situations, a rapid shift rotation where different shifts are worked every 2–3 days may be required. This can be caused by covering for other workers taking time off work or unusual work requirements such as changes in production or maintenance operations. This type of schedule tends to be more populated in Europe than the United States. Rapid shift rotations may reduce disruption to circadian rhythms because the body does not have time to adjust to a different wake/sleep schedule. The impact on social and family interactions can be less consequential. The vital concern is sleep deprivation. Quick rotations can result in large sleep debt and a strong sleep drive which can increase human fatigue risks. In addition, employees over the age of 45 may be less tolerant of rotating shifts whereas younger workers seem less affected by the circadian shift.

Direction of shift rotation is one area where sleep experts clearly agree. They recommend that shifts rotate forward from day to afternoon to night because circadian rhythms adjust better when moving ahead rather than backwards. The worker's sleep drive and circadian rhythms will work together to make it easier to sleep as the work shift is rotated. The direction of shift rotation rarely affects production schedules, but can improve performance and worker satisfaction making it an easy recommendation to implement.

Start times often influence the amount of sleep debt that workers are compiling. Starting first shift at 5 or 6 am will likely result in shorter sleep and greater fatigue than starting the shift a few hours later. The social customs and family obligations may make it difficult for workers to get a full night's sleep with an early morning start time. The boost of energy that occurs before bedtime with the body's circadian rhythms can add to the difficulty of attempting to go to sleep an hour or two early when an early morning start is required. The availability of public transportation and employee's safety getting to work can be another consideration in determining shift start times.

Length of rest between shifts often affects the ability to get adequate restorative sleep. It is recommended that a rest period of at least 24 h should be there after each set of night shifts. The more consecutive nights worked, the more rest time should be allowed before the next rotation occurs. API's RP 755 provides specific details in this area [3].

Shift length is typically 8 h, however, compressed workweeks with extended workdays are becoming more common. While this increases the number of alternatives when scheduling there are significant advantages and disadvantages to consider [1].

- Positive considerations:
  - Generally appeals to employees
  - More days away from the job
  - Reduces commuting problems and costs for employees
  - Has fewer startup and warm-up periods
- Negative considerations:
  - Can require overtime pay
  - Decreases job performance due to long work hours
  - Increases worker fatigue
  - Increases tardiness and early departure from work
  - Increases absenteeism
  - Potentially decreases workplace safety
  - Difficult to schedule child care and social events during the workweek

Split shifts occur when a worker's day is divided into two or more shifts. These are common in distribution centers and other facilities where work demand is segmented. A worker in a distribution center might work 4 h to load trucks in the early morning and then another 4 h to unload trucks in the late afternoon. These types of shifts are unpopular with workers as they lengthen the workday. Split shifts often do not allow enough recovery time between shifts as the break between shifts is too short to return home and rest. Early starts coupled with the late end of split shifts can result in fatigue, ill health, and disruption of personal social life.

Breaks for a meal and two or more shorter breaks are common during shifts of 8 or more hours. Frequent short breaks can reduce fatigue, improve productivity, and reduce errors and accidents. Short 5–15 min breaks every few hours may help maintain performance and improve safety, particularly when the task is demanding or monotonous. Breaks taken away from the workstation tend to be more beneficial than those taken at the workstation. When possible allowing workers some choice in when they take their breaks may allow them to be used to reduce fatigue and improve alertness. For example, toward the end of an evening shift as fatigue begins to set in, a worker may take more frequent breaks to walk and improve their focus.

Alternative work schedules are common in some industries such as health-care. Examples include having extended work days of 10 or 12 h. Some businesses have used four 10-h days which provides a 3-day weekend. Others have gone to 9-h days with every other Friday off. These schedules have the advantage of longer blocks of time off. However, the long work hours result in increased fatigue which may have adverse effects on safety and productivity. The physical and mental load of the task should be considered when selecting the length of a work shift. Exposure to chemical or other hazards should also be considered because this type of shift increases the daily exposure to such hazards (see the work schedules text box).

WORK SCHEDULES

In May 2004 Current Population Survey (CPS) was conducted on American household. The results found that over 27 million full-time wage and salary workers had flexible work schedules that allowed them to vary the time they began or ended work. This is 27.5% of all full-time workers. The Bureau of Labor Statistics of the U.S. Department of Labor reported that the proportion that usually worked a shift other than a daytime schedule is 14.8 %, which remained close previous surveys. By shift worked the percentages are 4.7% evening shifts, 3.2% night shifts, and 2.5% rotating shifts. 16.7% of men and 12.4% of women work an alternative shift.

The proportion of workers on alternative shifts was highest in leisure and hospitality industries (38.3%), mining (31.9%), and transportation and utilities (27.9%). Most (54.6%) of those working an alternative shift did so because it was the "nature of the job." Other reasons for working a nondaytime sched-ule included "personal preference" (11.5%), "better arrangements for family or child care" (8.2%), "could not get any other job" (8.1%), and "better pay" (6.8%). Many of those who worked night and evening shifts chose such schedules due to personal preference (21.0% and 15.9%, respectively) or because these shifts facilitated better arrangements for family or child care (15.9% and 11.0%, respectively). The vast majority of those with rotating, split, and employer-arranged irregular schedules reported the "nature of the job" as the reason for working a nondaytime schedule.

http://www.bls.gov/news.release/flex.nr0.htm

Other considerations include providing time off at desirable times such as holidays or weekends whenever possible. A successful example of this is a medical unit that allows nurses to select to have off either on Christmas Day or New Year's Day. The nurses with young families tend to select to have Christmas with their families while other nurses are more interested in having New Year's Day off. By allowing them to select one or the other, the majority is happy to have gotten their preferred day off from work. The supervisor found little resistance to this scheduling approach and was able to meet almost everyone's first choice of days off.

Another vital consideration is to give shift workers a great deal of advanced notice of their work schedules. This allows time to coordinate schedules with friends and family members. Some organizations will lock a schedule in place for 3–6 months and then allow workers to bid on schedules using a seniority system. Simple, predictable schedules result in fewer problems for superiors.

Policies related to trading shifts should also be considered. Although these practices can provide flexibility to workers, they can introduce unforeseen problems. Workers seeking extra hours or overtime might be eager to take on additional work shift at the expense of sufficient sleep. Hours of service limitations may have been met in the original shift but if shift can be freely traded then the limitations can be unwittingly violated.

10.3 MANAGING WORK-SHIFT SCHEDULING

Establishing or modifying a work-shift scheduling system can be time consuming and challenging. Workers are often resistant to changes in their schedules. As with any changes some employees will benefit whereas others will consider proposed changes as detrimental to their situation. It is vital to include workers, supervisors, safety experts, and human resources representatives in discussions and gain their agreement before making any modifications changes to working arrangements and/or schedules. As with any major policy change, it is important to clearly communicate information to those affected. Improving shift-work procedures may financially be beneficial by:

- Lowering absenteeism
- Decreasing staff turnover
- Reducing fatigue-related accidents and incidents
- Increasing efficiency
- Improving production quality
- Reducing medical costs

10.4 EVALUATING WORK SHIFTS USING THE HSE FATIGUE INDEX

The Health and Safety Executive (HSE), the UK's safety regulatory agency, in collaboration with WS Atkins Ltd. developed a method to calculate a fatigue index (FI) [4]. The index can be used to evaluate various work-shift schedules for the likelihood of causing fatigue. The method uses five factors:

Table 10.2 F1 Time of Day Score [4]

Start Time	0:00	1:00	2:00	3:00	4:00	5:00	6:00	7:00
Score	13	13	13	12	12	11	9	7
Start time	**8:00**	**9:00**	**10:00**	**11:00**	**12:00**	**13:00**	**14:00**	**15:00**
Score	4	2	1	1	1	2	3	4
Start time	**16:00**	**17:00**	**18:00**	**19:00**	**20:00**	**21:00**	**22:00**	**23:00**
Score	5	7	10	11	12	13	14	14

time of day (F1), shift duration (F2), rest periods (F3), breaks (F4), and cumulative fatigue (F5). The five factors in the FI are scored and then combined for a total score. These FI scores can then be used to compare various work schedule alternatives. Red zones indicating particularly troubling issues related to fatigue levels are also included in the FI system. Factors F1–F4 assess short-term fatigue and F5 assesses cumulative fatigue. The scores for F1 are shown in Table 10.2. These scores are for low levels of workload where work is routine and performed with little pressure. A value of 4 should be added to the scores in the table for complex task or those with strong time constraints.

The scores for F2-shift duration range from 0 to 22 and are based on start time and shift duration. As portion of the scores are in Table 10.3. The entire table can be found in the HSE document. When the shift is less than 8 h, then the score is prorated. For example, a 6-h shift is 75% or the 8-h score and a 4-h shift is 50%. If the work is higher in pressure or complexity the F2 score is increased by 30%.

10.4.1 **Fatigue Index Factor 3—Rest periods**

The third part of the FI score is associated with the timing and duration of rest periods prior to a shift. It is called a rest period score (RSP) and is based on the time of day that the rest period ends. Values are given in Table 10.4. A score of 0 is given if the rest period is longer than these limits. The RSP

Table 10.3 Selected F2 Shift Duration Score Values

Start Time	8 h	9 h	10 h	11 h	12 h	13 h	14 h	15 h	16 h
0:00	0	2	5	7	10	13	16	19	22
4:00	0	2	4	6	9	11	14	16	19
8:00	0	0	1	2	3	4	5	7	9
12:00	0	1	2	3	4	6	8	11	13
16:00	0	2	4	6	9	12	15	17	20
20:00	0	3	5	8	11	14	17	20	23

Table 10.4 Rest Period Score Values Used to Calculate the F3 Score

End of Rest Period	Rest Period Score (RSP)
04:00–08:00	2 for every hour by which the rest period is shorter than 13 h
09:00–12:00	2.6 for every hour by which the rest period is shorter than 12 h
13:00–18:00	3.5 for every hour by which the rest period is shorter than 13 h
19:00	2.2 for every hour by which the rest period is shorter than 15 h
20:00	1.8 for every hour by which the rest period is shorter than 17 h
21:00	1.5 for every hour by which the rest period is shorter than 19 h
22:00–03:00	1 for every hour by which the rest period is shorter than 20 h

score is multiplied by 10 plus the number of hours by which the following shift exceeds 8 h. For example, for an 11-h shift, the RSP score is multiplied by 11 + 2 = 13. The number obtained by this procedure is then divided by 20, and rounded to the nearest whole number to give the F3 score [4].

10.4.2 Fatigue Index Factor 4—Breaks

The fourth part of the FI score considers the time in which people are involved in work that demands continuous attention. The break factor is scored for work in which lapses of attention could increase the risk of an accident such as driving or monitoring a screen. If these do not apply, a score of 0 is given. The first step in calculating the F4 value is to divide the shift into four subperiods: 6:00–14:00 (morning), 14:00–17:00 (afternoon), 17:00–01:00 (evening), and 01:00–06:00 (night). If periods of sustained attention sometimes exceed 120 min for morning or evening periods, 60 min for afternoon, or 30 min for night without at least a 15 min break then the scores in Table 10.5 apply.

Table 10.5 F4 Score Values for Breaks

Typical Duration	Morning/Evening	Afternoon	Night
30–60 min	0	0	0.5
60–120 min	0	0.5	1
120–180 min	0.5	1	1
180–240 min	0.5	1	1.5
>240 min	1	1	1.5

Table 10.6 F5 Cumulative Fatigue Scores

Categories	1	2	3	4	5	6
Number of nights (N) in sequence	5	4	3	2	1	0
Number of earlies (E) in sequence	3	2	2	1	1	0
Number of lates (L) in sequence	3	2	2	1	1	0
Number of days (D) in sequence	0	0	0	0	0	0
Number of days off (DO) in sequence	−6	−4	−4	−4	−4	−4

10.4.3   **Fatigue Index Factor 5—Cumulative fatigue**

This score is considering the cumulative effect of fatigue (Table 10.6). Each 24-h period, starting at midnight, is assigned to a category according to the shift time whose mid-point is within that period. The periods are

Night (N): The part of the shift between 02:30–04:30
Early (E): The shift starts between 04:30–07:00
Late (L): The shift ends between 00:00–02:30
Day (D): Any other shift
Day off (DO): No shift during that period

If two consecutive shifts are in the consecutive 24-h periods and they are separated by a rest period of at least 30 h the score assigned to the second 24-h period should be reduced by 2. A cumulative fatigue score is calculated for each 24-h period by summing the CF scores for each day. If the CF score exceeds 15 then it is reduced to 15. If it is less than 0, then a value of 0 should be used. The F5 score for each shift is the CF score for the day on which that shift occurs. The F5 scores are then totaled to get a cumulative fatigue score. Days off work are excluded in this calculation.

10.5   **AN EXAMPLE OF HEALTH AND SAFETY EXECUTIVE'S FATIGUE INDEX**

Let's consider two very different work shifts. The first is four 10-h days, followed by 3 days off. The last 2 h are worked without an additional break. The score for this work schedule is 24 as shown in Table 10.7.

By comparison, consider the schedule in Table 10.8. Only 38 h are worked in this schedule but they are of varying lengths and various start time. The FI score for this much more varied schedule is 67.

A closer look reveals the reason for the differences in scores. In the first schedule, the F1 is relatively low for the 40 h/week schedule since all of the hours are worked during the daytime. In the second schedule, the F1 score

Table 10.7 HSE's FI score for a 10-h, 4 days/week schedule

	Day 1	Day 2	Day 3	Day 4	Day 5	Day 6	Day 7	Total
Shift	8:00–18:00	8:00–18:00	8:00–18:00	8:00–18:00	Off	Off	Off	
F1 score	4	4	4	4				16
F2 score	1	1	1	1				4
RSP score	0	0	0	0				
RSP total	0	0	0	0				
F3 score	0	0	0	0				0
Subperiods								
M or E	7	7	7	7				
A	3	3	3	3				
N	0	0	0	0				
F4 score	1	1	1	1				4
Shift type	D	D	D	D	DO	DO	DO	
Score	0	0	0	0	−6	−6	−6	
CF score	0	0	0	0				
F5 score	0	0	0	0				
Total score	**6**	**6**	**6**	**6**				**24**

Table 10.8 HSE's FI score for a 38-h/week variable schedule

	Day 1	Day 2	Day 3	Day 4	Day 5	Day 6	Day 7	Total
Shift	7:00-15:00	4:00-15:00	15:00-2:00	Off	21:30-5:30	Off	Off	
F1 score	7	11	4		13			35
F2 score	0	4	6		0			10
RSP score	0	0	0		0			
RSP total	0	0	0		0			
F3 score	0	0	0		0			0
Subperiods								
M or E	7	8	8		4			
A	1	1	2		0			
N	0	1	1		4			
F4 score	0.5	1.5	2.0		4.0			8
Shift type	D	E	L	O	N	O	O	
Score	0	3	3	−6	5	−6	−4	
CF score	0	3	6	0	5	0	0	
F5 score	0	3	6		5			14
Total score	**7.5**	**19.5**	**18**		**22**			**67**

is higher because day 2 and 5 are not worked during first shift. A higher F1 score is due to the increased risk of workers on the night or evening shift. The F2 value is also higher in the second schedule for the 2 days [3,4] that are 11-h days. These two shifts have different values for F2 because they start at different times of the day. The longer the shift the higher the F2 value. Research has shown that the rate of human errors increases the longer the shift. If a longer shift starts later in the day, this is compounded by fatigue and a higher risk score results. Both schedules have an F3 value of zero which indicates that adequate rest time is provided prior to each shift. The 40-h schedule has F4 values of 1 because a third break is not provided for workers in the last 2 h of the 10-h shift. The other schedule also has a penalty in the F4 score due to the limited number of breaks. The values vary by day due to the difference in the length of time without a break and the time of day.

F5 is the only cumulative value in the FI. In the first schedule the four shifts are all day shifts. The values come from Table 10.6. In the second 38-h schedule, the day 1 shift is considered a day shift. The day 2 shift is an early shift and day 3 is a late shift. Day 4 is an off day and day 5 is a night shift. The cumulative F5 values start at 0 for day 1, since it is a day shift. Day 2 is that 0 plus 3 for the first shift classified as early; this results in a value of 3 for day 2. Day 3 is the cumulative 3 plus 3 for the first late shift for a cumulative value of 6. Day 4 get a −6 since it is an off day which brings the cumulative value to 0. The day 5 value for the first night shift is 5 and it is added to the cumulative 0 for a total of 5. This factor in the fatigue index captures recovery time and variability.

The fatigue index values for the two schedules (24 and 67) provide a simple metric to compare the two alternatives. At first glance, a schedule for 40 h a week versus one for 38 h a week may not seem much different. However, as the difference in the FI value denotes there is a marked difference in the level of human factor that is likely to occur and an increased risk for the second schedule. The five factors in the index provide insight into potential problem areas for each schedule. The FI is an initial screening tool. It is intended to provide a comparative evaluation of the impact on fatigue for different shift patterns rather than an absolute measure.

REFERENCES

[1] Kroemer KHE, Kroemer HJ, Kroemer-Elbert KE. Engineering physiology bases of human factors/ergonomics. 4th ed. New York: Van Nostrand Reinhold; 2010.

[2] Rosa R, Colligan M. NIOSH Plain Language about Shiftwork. Available from: http://www.ccohs.ca/oshanswers/ergonomics/shiftwrk.html.

[3] A PI RP 755—Fatigue Prevention Guidelines for the Refining and Petrochemical Industries. http://www.api.org/Environment-Health-and-Safety/Process-Safety/Process-safety-standards/Standard-RP-755.

[4] Validation and Development of a Method for Assessing the Risks Arising from Mental Fatigue, Health & Safety Executive, Contract Research Report 254/1999. Available from: http://www.hse.gov.uk/research/crr_pdf/1999/crr99254.pdf.

Chapter 11

Work environment

11.1 INTRODUCTION

Human factors use scientific knowledge and principles to improve human interactions with devices, systems, and procedures. It can improve productivity, reduce injury risks, and increase user satisfaction. Human factors should be applied throughout the lifecycle of a product from the design stage, through operation, and ending with disposal. Preferably human factors should be used in the day-to-day operations of every facility and the design of every system. Many useful resources related to the application of human factors exist including the *Handbook of Human Factors and Ergonomics* [1]. An industry specific text is the *Human Factors Methods for Improving Performance in the Process Industries* [2]. It addresses design of control systems, human–computer interactions, management systems, and worker considerations.

Human factors can be justified by the benefits gained with respect to the following:

- Fewer accidents and near misses
- Improved quality
- Better worker satisfaction
- Reduced numbers of injuries and cumulative trauma disorders
- Improved productivity and efficiency

Although all of these are common benefits from making improvements based on human factors, it is often difficult to justify the cost of changes to improve the human factors of a process or equipment. Typically it can be virtually impossible to calculate numbers and the associated cost for these improvements such as how much an accident would cost and how many can be prevented. Human factors specialists often rely on productivity and efficiency increases to make the business case for changes even though it will benefit the organization in multiple ways.

Critics of human factors may hold the misconception that it is not a scientific field. The field has been around since World War II. It is a broad field that combines psychology, industrial engineering, and a number of other fields. The multidisciplinary nature may cause some to incorrectly assume that it is ill defined, which is not the case. However, there is a major challenge for human factors, which is humans themselves. People vary, people often do unpredictable things, and there are ethical limits to the testing we can do on human subjects. Despite these limitations, there are still clear recommendations and guidelines that can be applied to minimize the risk of fatigue and fatigue-related accidents on the job. This chapter will focus on environmental factors that can affect fatigue, sleepiness, and arousal in industrial applications.

11.2 LIGHTING

Much research has been performed on lighting and human performance. There are industrial standards from the Illuminating Engineering Society of North America (IESNA) and American National Standards Institute (ANSI) for numerous lighting applications. Various groups have developed recommendations on the basis of different concerns. Astronomers advocate for lighting that minimizes the problem of light pollution, the degradation of the night sky due to increased artificial lighting. Environmentalists support lighting that reduces the energy and environmental pollution. Others advocate for greater lighting levels to improve safety and security in some applications. In this section, we will consider how light influences human performance and the body's circadian rhythms.

Two characteristics of lighting are considered for their effect on human factors, quantity and quality of light. The quantity of light needed varies with the application. It is measured as illuminance, the amount of light falling on a work surface, and is measured in the unit lux internationally and foot-candles in the United States. Table 11.1 lists selected lighting levels recommended for US government facilities and can provide a rough guideline of common lighting levels for differing facilities [3]. Glare is an example of light quality. It can be direct from a light source to the individual or indirect glare. Sunlight from a window that bounces off a computer screen

Table 11.1 Selected Recommended Lighting Levels from the GSA

Application	Lighting Levels
Internal corridors, stairwells, lobbies, dining areas, loading docks, mechanical rooms	200 lux
Conference rooms	300 lux
Training rooms, offices, structured parking entrances, physical fitness space	500 lux

and hampers an individual's vision is an example of indirect glare. Glare can be managed by adjusting lighting levels, moving computer screens, closing window blinds, and moving light sources or workers. In some settings, lower ambient lighting can increase operator comfort and reduce screen glare and eye strain. The concern for fatigue risk management is whether lower ambient lighting will decrease alertness and increase the chance of workers nodding off to sleep.

Biological research has shown that light has important nonvisual biological effects on the human body. Good lighting has a positive influence on health, well-being, alertness, and even on sleep quality [4]. Typically morning sunlight synchronizes our internal body clock to the 24-h day [7]. This works well for the circadian rhythms of day-shift workers (see chapter Biology of Sleep). Workers on evening and night shifts have lighting cues that can add to the desynchronization of their circadian rhythms. Not only the time of the lighting but also the amount of lighting can influence the health, well-being, and alertness of workers. van Bommel [4] summarizes studies with different lighting levels and high lighting levels for night workers. In one study, he found improving lighting levels from 300 to 500 lux increased productivity 8%. A similar change from 300 to 2000 lux had a 20% productivity improvement. van Bommel also reported that higher lighting level resulted in significantly increased arousal levels, more alertness, and better moods. Rea et al. [6] found that quantity, spectrum, spatial distribution, timing, and duration are key factors when studying the practical effects of lighting on humans. They found it can influence depression, sleep quality, alertness, and health. Unfortunately the research is not to the level to provide clear application guidelines to lighting practitioners or managers.

11.3 **TEMPERATURE**

The human body tries to regulate itself to maintain a relatively stable core temperature in the range of 97–99°F. If the body deviates from this narrow range major health consequences, even death, can occur. Blood circulation can be increased or decreased by the body in an attempt to maintain this healthy temperature range. Heat is exchanged between the human body and the environment; this occurs through conduction, convention, evaporation, and radiation [7]. As the body's temperature moves away from the desired level, our performance suffers. This can result in increased errors, fatigue, and exhaustion; and in extreme cases death.

Temperature can be a concern for fatigue risk management. One function of our body's circadian rhythms is to regulate body temperature. Room temperature can make it more conducive to fall asleep (see chapter: Sleep

Hygiene Recommendations). Physical exertion and the associated heat production can influence sleepiness and performance in sleep-deprived workers. A research study [8] found that physical exercise significantly reduced workers' feelings of sleepiness depending on the magnitude of the core body temperature elevation. This finding indicates that suppressing heat loss could help prevent sleepiness during the night shifts and avoid decreased performance levels.

Room temperature influences perceived comfort. As the temperature deviates from a comfortable level, our sensation of discomfort goes from annoyance to pain. This is a signal from our body to take corrective actions whether it is to adjust our activity level, move to a new location, adjust our clothing, or manipulate the heating/air conditioning system. Overheating leads to weariness and sleepiness. Overcooling leads to restlessness, which can reduce alertness. There are four climate factors [7] that impact comfort:

- Air temperature
- Air humidity
- Air movement
- Temperature of adjacent surfaces

As with many areas of human factors, there is human variability. Some have speculated that men and women have different internal thermostats; though this may lead to interesting discussions in couple's therapy, the scientific research is inconclusive. Perceived differences may be swayed more by clothing and body mass than gender.

11.4 **NOISE**

Noise can be defined as an unwanted sound. It is a concern in work settings for a variety of reasons. Sound can be used to convey information such as warning alarms or public address systems. Exposure to moderate or high levels of noise can lead to hearing loss over time. Noise can provide mental stimulation or distraction. Safety guidelines and hearing conservation programs are common in industrial settings. In the United States, OSHA has regulations in this area. Exposure is limited on the basis of loudness of the sounds (measured in decibels) and duration of exposure per day. A hearing conservation program requires monitoring noise exposure levels for employees exposed to noise at or above 85 decibels (dB) averaged over 8 working hours.

Noise above these levels can be managed in three ways. First, the noise can be reduced or eliminated. This can be done in a variety of ways including enclosing noisy machinery, preventing vibration, replacing worn equipment belts or loud equipment with newer quieter models. The second method is to provide personnel protective equipment (PPE) such as ear plugs or ear muffs. These have noise reduction ratings (NRR) which measure the potential to reduce the noise exposure for the person wearing the PPE. Of course if workers do not wear the PPE, nothing is gained. The third approach is to limit the exposure time by job rotation. This reduces noise exposure in the short run but cumulative hearing loss can still occur over time.

With respect to fatigue risk management, we are considering the psychological effects of noise more than the physiological effects. In pervious chapters, white noise was discussed as an aid in falling asleep. White noise is a meaningless, unobtrusive sound that can mask other sounds that might distract something from falling asleep. There are people who pay money for machines that mimic the sound of rain falling or the ocean. Some work settings have equipment with motors that generate similar steady sounds that can encourage unwanted sleep in tired workers.

Noise can be positively stimulating in the right circumstances. Music, conversation, or other wanted sounds can reduce boredom and monotony, and, in some settings, improve mental performance and concentration. In general, individuals working on mentally demanding tasks typically perform better with less noise in the environment. Many people tend to turn off surrounding sounds from radios, TVs, and other sources when they need to concentrate. Some people advocate the use of noise, such as loud rock music to keep themselves awake, particularly while driving. A former student swore that the secret to staying awake on a 2-h commute late at night was listening to political talk radio from an opposing point of view. Unfortunately, there is little research to support these beliefs. A sleep-deprived worker struggles to remain awake regardless of whether the radio is on or not.

SLEEPY DRIVERS AND THE CAR'S ENVIRONMENT

Researchers in Sweden investigated the effects of opening the vehicle's window and listening to music as countermeasures against driver sleepiness. They used both a subjective measurement of sleepiness and physical measurements of eye blinks to evaluate whether either of these two common driving tricks worked to keep sleepy drivers more alert. Opening the window had no effect on sleepiness. There was a minimal positive effect from listening to music. However, individual differences were large. The difference whether it was daytime and nighttime was much greater for sleepy drivers than having the radio playing or not. Other studies have shown that drinking caffeine and/or stopping for a nap to be more effective countermeasures against tired driving than open windows or music.

11.5 **VIBRATION**

Vibration is the mechanical oscillation about a fixed reference point. The study of human response to vibration is a multidisciplinary topic that includes biology, psychology, biomechanics, and engineering. It is typically classified as whole-body vibration, hand-transmitted vibration, or motion sickness. These typically involve frequencies from 1 to 100 Hz for whole-body vibration and from 8 to 1000 Hz for hand-transmitted [9]. The automobile industry has conducted significant research on whole-body vibration to improve driver and passengers' comfort while riding in vehicles. Hand-transmitted vibration is a common concern with power tools.

As with many environmental factors, vibration affects can range from an annoyance to a causing injuries. Table 11.2 is a summary of potential effects from human vibration [1,2,7,9].

The effect of vibration on health and human performance has been studied for a number of years and clear results have been found. From this we can conclude that avoiding vibration when possible is desirable. However, much less research has been conducted on how vibration may influence sleep, fatigue, or alertness. One study [10] experimentally investigated sleep disturbance caused by nocturnal vibration. They found a negative impact on sleep from vibration and that it increased with greater amplitudes. We cannot make any recommendations in this area related to human fatigue risk management, nevertheless substantial levels of human vibration should not be ignored.

Table 11.2 Effects of Human Vibration

Effect Type	Common Effects
Physiological effects	■ Impaired psychomotor performance ■ Decreased visual perception ■ Increased energy consumption
Psychological effects	■ Impaired visual performance ■ Decreased mental processing of information ■ Diminished task performance
Health effects	■ Damage to spine ■ Increased risk of cumulative trauma injuries (such as Carpal Tunnel Syndrome) ■ Increased nausea and vomiting ■ Interference with breathing ■ Increased occurrence of Raynaud's Syndrome (damage to fingers) ■ Muscular tension, headaches, and general discomfort

11.6 **COLOR**

There are psychological effects of color such as changes in perception or arousal. These may be caused by prior subconscious associations, such as red means stop, or hereditary factors. They can influence human performance and human behavior. Some designers have realized the benefit of using particular colors to achieve desired behaviors. Prisons and psychiatric hospitals often have a room with the walls painted pink. Researcher found a color called Baker-Miller pink can have a calming effect and reduce hostility, violence, and aggressive behavior. These rooms are used to calm patients or prisoners that are out of control.

In general, researchers have found blue, green, and brown tend to be restful colors, whereas red, orange, and yellow are considered exciting or stimulating [7]. Interior designers use color to establish a "feel" to a space. There are websites that tout the psychological effect of colors in a restaurant. They claim that a red table cloth will make people eat more. Yellow is good for fast-food restaurants because it motivates people leave quickly. Green is relaxing and will make people more comfortable. All of these recommendations should be taken with some degree of skepticism. Color effects for wall paint do not necessary carryover to other applications. Contrast, lighting, and textures, in addition to the wide variety of colors; complicate psychological studies in this area. In addition, the human mind is very complex and there is a wide spectrum of factors that influence human perception, thus it is difficult to make general recommendations in this area. However, there could be some measurable effect of color on a person's level of alertness in some applications which might warrant further consideration.

REFERENCES

[1] Salvendy G. Handbook of human factors and ergonomics. 4th ed. Hoboken, New Jersey: John Wiley & Sons, Inc.; 2012.

[2] Crowl D, editor. Human factors methods for improving performance in the process industries, CCPS. New York: Wiley-Interscience; 2007.

[3] http://www.gsa.gov/portal/content/101308 2003 Facilities Standards, 6.15 Lighting. U.S. General Services Administration.

[4] van Bommel W. Non-visual biological effect of lighting and the practical meaning for lighting for work. Appl Ergonom 2006;37:461–6.

[5] Rea MS, Figueiro MG, Bullough JD. Circadian photobiology: an emerging framework for lighting practice and research. Lighting Res Technol 2002;34(3):177–90.

[6] Boyce, R, Beckstead, J. Eklund H. Lighting research for interiors: the beginning of the end or the end of the beginning. Lighting and Research Technology, Dec 2004;36,(4):283–293.

[7] Kroemer K, Grandjean E. 5th ed. Fitting the task to the human. 5th ed. London: Taylor & Francis; 2001.

[8] Yasuhiro M, Kazuo M, Kohtoku S, Tetsuo S, Yasuo H. Physical activity increases the dissociation between subjective sleepiness and objective performance levels during extended wakefulness in human. Neurosci Lett 2002;326:133–6.

[9] Mansfield N. Human response to vibration. Florida: CRC Press; 2005.

[10] Smith MG, Croy I, Ögren M, Persson Waye K. On the influence of freight trains on humans: a laboratory investigation of the impact of nocturnal low frequency vibration and noise on sleep and heart rate. PLoS ONE 2013;8(2):e55829.

Work task design

12.1 INTRODUCTION TO WORK DESIGN

Work task design has a long history in the field of industrial engineering (IE). Over the last 125 years IEs have been analyzing work for a variety of reasons. Around the turn of the 20th century Frederick W. Taylor led the effort to establish time standards using work measurement techniques [1]. Taylor proposed that the work of each worker should be planned by management. He felt that productivity could be improved through the application of scientific principles to design the work method and tools. Taylor did not respect workers and assumed they would slack off as much as allowed. Early people involved with work task design tended to follow in this vane and were motivated by improving output and profits. They were often referred to as efficiency experts.

Around the same time period, Frank and Lillian Gilbreth founded the field of motion study. They studied the body motions used in performing a task. Unnecessary motions were then eliminated and necessary ones were simplified to maximize efficiencies. Their focus was more on the workers and improving the process. They worked to design tasks that could be performed by disabled soldiers who had returned from World War I. The Gilbreth's work also resulted in cost savings. Bricklaying is a task that has been around since the time of the pyramids. A satisfactory rate in the Gilbreth's day was considered to be 120 bricks per hour. After studying the task, they were able to make simple changes that increased the average number of bricks laid to 350 per hour [1]. By reducing the number of bending motions involved, their more efficient method was physically easier on the bricklayer in addition to more productive.

During the following decades, the field of work design became better defined. Industrial psychologists made important contributions by expanding the field to include cognitive aspects of work. One early study was the Hawthorne Study at Western Electric where illumination levels were adjusted to determine the effect on workers' performance. As the lighting was increased, the productivity improved; when the original lightbulbs were reinstalled, the productivity improved again. A realization was made that the research study was poorly controlled and that worker motivation played a pivotal role in the results [1]. The performance improvements were due to the attention paid to the workers participating in the study and the notoriety of their involvement rather lighting levels. This study and other related ones demonstrated the importance of worker perception, motivation, and feedback in work settings.

In the 1970s and 1980s, the US government responded to expensive cost overruns among Department of Defense (DoD) contractors by adopting a military standard (MIL-STD 1567A). This standard required military equipment contractors and subcontractors to follow government regulations for work design and time standards. The goal was to control cost overruns by requiring clear work plans with time requirements (and associated cost bids) on the basis of accurate time values determined by well-established work measurement techniques. Some benefits were achieved but the military standard was canceled in 1995 [1].

With the growth of the personal computer in the 1980s, work design concepts were applied to our dealings with both hardware and software. The field of human–computer interaction (HCI) was born. The classic cautionary HCI tale for engineers is the Three Mile Island Accident. There were small radioactive releases during the incident, but no detectable health effects on plant workers or the public [2]. Since the accident events in the control room have been studied to improve human–computer interactions in complex monitoring situations. The work continues by the Abnormal Situation Management Consortium (http://www.asmconsortium.net) which is composed of leading processing companies and universities working together to improve safety in the processing industry.

ASM CONSORTIUM GUIDELINES

The ASM Consortium has developed guidelines describing best practice for the processing industry. They are based on research at member companies and represent best practice applying human-centered design to the management and operation of plants. Publications available for purchase include the following:

- Effective console operator HMI design practices
- Effective alarm management practices
- Effective procedural practices

OSHA's interest in work design, particularly ergonomics, has varied over time. An ergonomic regulation covering most industries was passed and

Table 12.1 Effects of Fatigue	
Mental fatigue	■ Slower reaction times
	■ Lapses in attention and vigilance
	■ Impaired judgment and decision making
	■ Loss of motivation
Physical fatigue	■ Feeling tired and/or sore
	■ Temporary decrease in muscle strength
	■ Decrease in physical performance
	■ Decline in cognitive performance

scheduled to take effect in January 2001. Shortly after taking office President George W. Bush signed a resolution resending the regulation [3]. Since that time OSHA has taken a piecemeal approach to regulations in this area. Certain industries with high levels of cumulative trauma injuries have been specifically targeted for regulations. OSHA has also written fines under the General Duty Cause which calls on employers to provide a safe, hazard-free workplace for American workers.

Although good work task design commonly has positive benefits in improved safety, decreased cumulative trauma injuries, enhanced productivity, and evaluated worker satisfaction; the remainder of this chapter will focus on the work task design and fatigue risk management. How the mental aspects of a task can influence monotony or stimulate workers are vital concerns in fatigue risk management because they can directly affect the worker's level of alertness. In a similar fashion, physical aspects of the job can have a marked effect on workers' level of fatigue [4]. Table 12.1 summarizes the effects of both mental and physical fatigue.

12.2   **WORK STRESS**

Stress is a person's reaction to a threatening situation and it triggers a biological reaction. In our modern world, stress is inescapable. Workplace stress has been studied by many researchers. Atwood et al. [5] summarizes the sources of stress; a modified version is presented in Table 12.2. Their book is a good source of more detailed information on the topic of work design geared specifically toward the process industry.

For over 100 years, researchers have investigated how characteristics of a task relate to performance. Two early pioneers were Yerkes and Dodson who in 1908 presented a law that bears their names [6]. This was based on research with rats exploring the relationship between pain stimulation and performance for solving a maze with varying difficulties. They proposed

Table 12.2 Sources of Work-Related Stress

Workplace Environment	Physical Environment	Individual Differences
Contributing factors:	Contributing factors:	Contributing factors:
■ Lack of control	■ Noise levels	■ Personality
■ Lack of management support	■ Poor lighting	■ Lack of experience
■ Work overload	■ Vibration	■ Lack of skill or knowledge
■ Excessive task demands	■ Temperature extremes	■ Individual health
■ Job uncertainty	■ Air quality	■ Life balance issues
■ Lack of communication		
■ Job insecurity		
■ Work relationships		
Results in:	Results in:	Results in:
■ Increased stress	■ Increased stress	■ Increased stress
■ Health complications	■ Greater errors	■ Poor performance
■ Employee turnover	■ Health complications	■ Job dissatisfaction
■ Poor performance	■ Poor performance	
■ Job dissatisfaction		

an inverted U-shaped function to define the relationship between arousal and performance. Over the years others have expanded this law to address the effect of motivation, stress, and other factors on performance measures including learning, problem-solving, and memory. Fig. 12.1 illustrates the law. Low levels of stress or arousal result in weak performance, as do high levels. A moderate level of stress or arousal will result in the best performance from the individual.

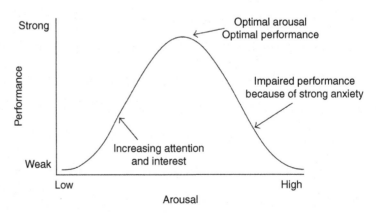

■ **FIGURE 12.1 Yerkes–Dodson Law.**

Sports can provide a useful demonstration of the law. Imagine an Olympic athlete. After years of training, her daily workouts have become commonplace with little stress or stimulation. This would place the athlete on the left-hand side of the curve. The thrill and excitement of trying out for her nation's Olympic team certainly increases her stress and motivation; she moves to the middle of the curve. With the added incentive, she sets a personal record and makes the team. Yerkes and Dodson would consider this an optimal performance at the top of the curve. Later at the Olympic with the eyes of the world on her, the pressure and stress are too much for her. She makes errors and does not achieve the level of performance she had before; placing her on the right-hand side of the curve.

This is a nice illustration but some athletes have the performance of their lifetime under the stress of the Olympics. For them the stress of that once-in-a-lifetime event places them at the top of the curve with a peak performance. What is different? The individual. What is the optimal level of stress? We do not know. No one has been able to define it or measure it. Humans are complex creatures; many factors can influence the shape of the curve and an individual's position on the curve. The level of training, experience, and motivation can influence someone's position on the curve. Personal factors such as genetics and psychological tolerance to stress can influence this as well. One's level of fatigue and their circadian rhythms can also impact performance. The nature of the task such as mentally complex versus physical demand can alter the shape the curve. The effects of stress can be grouped into three categories [7].

- Psychological: stress can produce emotions such as frustration and cause a drive to take action.
- Physiological: stress can have short-term effects such as increased heart rate and long-term health effects such as heart disease, ulcers, and other illnesses.
- Performance: stress can impede mental performance by decreasing our ability to process information and make decisions.

The consequences of mental overload on decision making have been documented to include decreases in accuracy, poorer information processing, decreased use of strategies, and locking onto a single solution or strategy. Cognitive tunneling is a term used to describe this human response. Some of these negative characteristics can be managed by training and design. A specialty within the field is alarm management. It strives to provide the worker with the necessary information quickly and in sufficient detail, while not causing an overload.

Surprisingly to some, workers who are underloaded often have the similar issues to those that are overloaded. Research has shown that workers

Table 12.3 Strategies for Coping with Stress

Categories	Coping Strategies
Work task	■ Address uncertainty and ambiguity in work tasks ■ Improved job training ■ Clearer work processes and procedures
Physical	■ Regular exercise ■ Meditation, breathing exercises, and other stress management approaches ■ Follow healthy lifestyle and nutritious diet
Mental stress	■ Take breaks or vacations ■ Set realistic expectations ■ Set priorities ■ Stress management training

attempting to sustain attention on a vigilance task, such as monitoring a control panel, in low-arousal environments can be just as fatigued as those in high-arousal environments. Their work performance can suffer just as their overworked counterparts' work did. The term vigilance decrement is used for the decrease in performance over time for individuals in monitoring tasks. These declines are measured in minutes rather than hours. Remediation methods can include the following [7]:

■ Short shifts with frequent rest breaks or changes in the task
■ Enhance warning or alarm signals to be more salient to the operator
■ Introduce false signals so detection is not so rare
■ Increase the level of stimulation in the work

There are different approaches to dealing with stress. Their effectiveness will vary with the individual and the situation. Table 12.3 presents a summary of stress-coping strategies modified from Atwood et al. [5].

12.3 ADMINISTRATIVE SOLUTIONS FOR WORK DESIGN ISSUES

After Taylor's push for job simplification and highly designed work practices, there has been some anti-Taylorism backlash. Critics feel that repetition of the same detailed process over and over again all day long has negative consequences for the workers. The job becomes very monotonous which can lead to poor work quality, a decrease in the worker's alertness, and high turnover rates among workers. Ergonomists are concerned that high repetition of motions will contribute to cumulative trauma disorders (CTDs) such as Carpal Tunnel Syndrome. CTD are injuries to the soft tissue, such as nerves, muscles and tendons, over time. A worker performing the same motion continuously in a physically demanding job is at higher risk for one of these injuries, the cost of which are often born by the employer.

One of the remedy for this problem is job rotation. With job rotation, a worker moves to different tasks during a work shift. An example of this is a transportation security administration (TSA) agent processing incoming airplane passengers. An agent might start his day monitoring the X-ray of carry on luggage. This is a very monotonous task and is mentally draining. If the agent switches to checking passengers' identification and boarding passes, it will give him a change. It is still a monitoring task, but the increased movement exchanging papers with passengers and interacting with them will provide a bit of a change and can decrease his mental fatigue and increase his alertness. Rotating to the job of moving plastic bins from the end of the carryon baggage screening conveyor to the front is a change of pace. He is now lifting and walking; physically exerting himself. The task of overseeing passengers during the body scan also provides a mental change and physical work. Rotating between these tasks is beneficial both ergonomically and cognitively. The agents get some periods walking and moving along with other periods of sitting down. This mixture reduces the risks of CTDs. The job rotation also benefits the mental work performance as the agents switch from the highly demanding monitoring task requiring sustained attention to the tasks that do not require such mental vigilance.

Well-designed job rotation changes the physical and mental demands on the worker. The changes should be throughout a work shift to gain the best benefit. Many mass-production facilities move workers at each break which typically provides a change every 2 h. Some facilities such as the TSA agents or casinos move employees much more frequently such as every 15 min. Workers typically enjoy the variety provided by job rotation. With more people able to do different jobs, management can gain from the increased flexibility in the workforce. If one of the tasks in a job rotation is particularly hazardous, labor unions may be resistant to implement job rotation. Their logic is that if a job such as loading raw materials is physically demanding and at a higher risk for a CTD or other injuries, then rotating people through this task is increasing the number of workers at risk for injury even though the likelihood is decreased for each due to the shorter exposure. In this situation, the solution is to redesign the job to eliminate the hazard.

Horizontal job enlargement has been suggested as a solution to repetitive work since the 1930s [8]. Horizontal job enlargement is expanding a job to include a greater number and variety of activities. It can be achieved by combining two or more differing jobs, increasing the length and range of tasks. In a manufacturing setting this could be achieved by having a machine operator also complete associated paperwork, restocking supplies, and cleaning the work area. In an office setting this might include having a receptionist also responsible for ordering supplies. The benefits to the worker can be similar to those of job rotation. Workers are doing more varied work. With respect to fatigue risk management, this can result in activities that are less monotonous and potentially increase the individual's alertness. Management has the increased training cost as in job rotation but may also have other ongoing costs. For example, the machinist may be paid more than the person who previously performed some of these tasks. Also the machine may sit idle while the machinist is doing these new added activities. In some settings the additional costs have been deemed acceptable. An example would be having a security guard walking rounds periodically in addition to monitoring security cameras. The benefits of a more alert guard due to physical extension and mental change of walking outweigh the costs. Both job enlargement and job rotation might benefit a FRMS in a chemical processing setting by reducing monotonous work.

12.4 **WORKPLACE EXERCISE**

Exercise can be added to jobs that are stationary and prone to boredom and fatigue. The National Institutes of Health (NIH) Office of Research Services provides exercises and stretches that can be performed at a workstation to reduce strain on the individual [9]. These can also improve alertness and provide short physical and mental breaks. Eye exercises can prevent headaches and provide a very brief mental break. Suggested eye exercises include the following:

- Blinking or yawning: It helps moisten and lubricate the eyes
- Expose eyes to natural light
- Close and rest your eyes
- Slow and gentle eye movement up and down; left and right
- Change focus distance from near to far

Musculoskeletal exercises, similar to those done before a workout, can be done in the workplace. These have both ergonomic and cognitive benefits. Suggested musculoskeletal exercises include the following:

- Neck stretches
- Shoulder shrugs
- Foot rotation
- Shaking hands and arms
- Wrist stretches
- Back stretches

Commercial software is available to monitor computer work and to suggest periodical stretching breaks for users. Many of these include detailed online instructions for stretch breaks. Many vendors sell training materials and handouts on this topic. Several free apps are also available to remind workers to take a break and suggest restorative activities to do during the break. Resistance to this form of software is often very strong. Workers can be apprehensive of being interrupted while they are working and think stretching at work will make them look silly. Involving them in the selection process and explaining the benefits are keys to successful implementation.

More general efforts can be used to increase movement during the workday. Incentives to reach a target such as 10,000 steps per day are becoming more common as a part of corporate wellness programs. Devices such as the Fitbit, along with apps for the smartphone, can be used to monitor and remind the wearer of the need for physical activity. Although the goal of these devises and programs is improving one's overall health, the increased activity can benefit alertness and provide both mental and physical breaks, as well as improve overall health.

WORKPLACE WELLNESS SCORECARD

The Centers for Disease Control and Prevention (CDC) has developed a Worksite Health ScoreCard (HSC). It is an online tool to help assess whether there is evidence-based health promotion interventions and strategies in the workplace. These can prevent heart disease, stroke, high blood pressure, diabetes, and obesity. The CDC realizes that America is facing an epidemic of chronic diseases that hurt business productivity and raise health-care costs. They believe that wellness programs can improve this.

More information on designing and evaluating a wellness program can be found at http://www.cdc.gov/healthscorecard/introduction.html

---

12.5 ENGINEERING SOLUTIONS FOR WORK DESIGN ISSUES

Compared to administrative solutions, engineering solutions redesign the work task or workplace to eliminate the problem rather than manage it. A wealth of research exists in the field of human factors on designing how humans receive information from systems and how they can adjust systems. The effectiveness of providing information can be subject to a diversity of factors. For starters, how is the operator's attention captured; alarms, warning light, or other means? Is the design loud enough, bright enough, or lost in the surroundings? Once it is detected, does the operator understand it? Everything from the wording of the message to the message length or the operator's expectations can affect the success of the comprehension. The operator needs to respond correctly. This is achieved by well-designed procedures, effective training, and good decision-making techniques. Typically the next step is responding with an adjustment to the system by the operator. This can be done by a variety of controls including knobs, pedals, button, or semi-automated protocol. As with most things, the usability of these is based on the design. In the realm of fatigue risk management it is important to remember that the workers' performance can be inhibited by fatigue.

We discussed changes to human performance due to fatigue in a prior chapter. As attention, vigilance, and reactions declines the engineering solution is often to make the signal to the worker bigger, louder, and clearer. What is sufficient to capture the attention of a well-rested worker might not be sufficient for a fatigued worker or one at a low point within the circadian rhythm. The sleep-deprived worker may take longer to react or make poor decision. The way to combat the effects of tired workers is to have clearly established emergency response and over train these. The negative impact on memory and cognitive function can be reduced by written guidelines, procedures, or checklists. All of which reduce the mental demand on the tired worker. Engineering the job and the work environment for the potentially fatigued worker will benefit others as well. Distracted workers, novice employees, and mediocre personnel may have very similar limitations in their performance. Enhancements made in response to a FRMS, can have payback in other situations as well.

12.6 ERROR PROOFING

The poet Alexander Pope is best known for his quote that "to err is human". This is true and a sleepy, fatigued worker makes even more errors that a well-rested counterpart. Human error needs to be considered by those working in safety. How can we prevent human error? How can we minimize the impact of human error? These are questions that should be asked by safety professionals in the process industry. The answer is often in how the job or equipment is designed.

An unfortunate example of this in the process industry can be found in the Chemical Safety Board's (CSB) investigation into a fatal accident at the at the Formosa Plastics' Vinyl Chloride Explosion in Illioplois, Illinois in 2004 [10]. An operator was overseeing a production process manufacturing PVC (polyvinyl chloride). The operator intended to open a valve on reactor vessel during a cleaning operation. He mistakenly opened the valve on a different reactor that was full, releasing highly flammable contents. The two reactor tanks were not clearly labeled and no warnings were provided in the written instructions about correctly identifying the vessels. They did not have display lights to inform the workers of the reactor vessel status which could have helped the worker verify he was working on the proper vessel.

A process hazard analysis (PHA) is a tool to evaluate and protect against human error. During a PHA-specific process, hazard scenarios, including human error, and the potential consequences are identified. These hazards should be addressed by safeguards to reduce the risk or mitigate the consequences. OSHA's PSM standard (29 CFR 1910.129) requires facilities using highly hazardous chemicals to conduct a PHA and revalidate it periodically. This had been done at the plant by previous owners, Borden Chemical. The CSB report [10] findings concluded:

1. *Borden Chemical did not adequately address the potential for human error: a) Borden Chemical did not implement 1992 process hazard analysis (PHA) recommendations to change the reactor bottom valve interlock bypass to reduce potential misuse. b) In a 1999 PHA, Borden identified severe consequences for opening the reactor bottom valve on an operating reactor, but accepted the interlock, controlled by procedures and training, as a suitable safeguard.*
2. *Formosa-IL did not adequately address the potential for human error: a) After a 2003 incident at FPC USA's Baton Rouge facility, Formosa-IL did not recognize that a similar incident could occur at the Illiopolis facility or take action to prevent it. b) Formosa-IL site management did not implement corrective actions identified in the investigation of a similar incident in February 2004 at Formosa-IL.*
3. *Formosa-IL relied on a written procedure to control a hazard with potentially catastrophic consequences.*

A great deal of research has been performed on preventing human error. The term poka-yoke comes from lean manufacturing and it means mistake proofing. Its goal is to eliminate defects by preventing, correcting, or signaling human error when it occurs. As an example, USB ports are designed with a shape that prevents people from inserting a connector incorrectly. A software example of correcting a human error would be software that changes the letter o to a zero in an input field that requires numbers. An interlock on a device that causes a machine to shut-off when accessed during operation is a mechanical example of an error correction. Adding warning messages or signal lights is an example using poka-yoke to signal a human error.

The American Society for Quality (ASQ) has many excellent resources on this topic. Table 12.4 summarizes the steps they recommend in mistake proofing [11]

Table 12.4 Mistake Proofing Procedure

Number	Mistake Proofing Steps
1	Obtain or create a flowchart of the process. Review each step, thinking about where and when human errors are likely to occur. A PHA can be useful for this step.
2	For each potential human error, work back through the process to find its source. What would be the root cause for the mistake?
3	For each error, think of potential ways to make it impossible for the error to occur. Consider the following: Elimination: Eliminating the step that causes the error. Replacement: Replacing the step with an error-proof one. Facilitation: Making the correct action far easier than the error.
4	If it is not possible to prevent the potential error to occur, think of ways to detect the error. Consider the following: Inspection: How and when will the error be detected? Parameter: What attribute will be inspected to detect the error?
5	If an error does occur, how will the worker be alerted? Warning bells, buzzers, lights and other sensory signals are common methods. Color-coding, shapes, symbols, and distinctive sounds help the worker understand the error message.
6	Control functions prevent the process from proceeding until an error is corrected or the condition which would allow an error is corrected. This can minimize the effect of an error.
7	After choosing the best mistake proofing method, test, implement, and evaluate it.

12.7 HUMAN RELIABILITY ANALYSIS

Human reliability analysis (HRA) quantitatively predicts and evaluates human performance in complex systems. It considers error likelihood, probability of task completion, and response time. The goal of HRA is identify high-risk areas and aid in reducing these risks. This type of analysis can be done to supplement PHA. Two terms are worth defining:

Human performance reliability: The probability that a human will perform a given task under specified conditions

Human reliability: The probability of a successful human performance on a specific task during system operations *with a given time limit*

There is a variety techniques that can be used to quantify these probabilities. *Kodak's Ergonomic Design for People at Work* [12] gives a good summary of them. It is also a very good general text on design workstations and procedures.

Example of common human reliability analysis techniques:

- Techniques for human error rate prediction (THERP)
- Success likelihood index methodology (SLIM)
- Human error assessment and reduction technique (HEART)
- Absolute probability judgment (APJ)

Even with these techniques human reliability is difficult to evaluate. Data rarely exists and expert judgment is used to establish probabilities. These can be subject to bias. Values can be conservative due to a bias "that can't happen here" or the desire to avoid the additional time and expense of addressing the risk. Values can be influenced by a liberal bias for "safety at all cost" or recent negative events.

REFERENCES

[1] Freivalds A. Neibel's methods, standards, and work design. 12th ed. Boston: McGraw Hill; 2003.

[2] http://www.nrc.gov/reading-rm/doc-collections/fact-sheets/3mile-isle.html.

[3] https://www.osha.gov/archive/ergonomics-standard/archive.html.

[4] Abd-Elfattaha HM, Abdelazeim FH, Elshennawy S. Physical and cognitive consequences of fatigue: a review. J Adv Res 2015; 6(3):351–358. Available from: http://www.sciencedirect.com/science/article/pii/S2090123215000235.

[5] Atwood D, Deeb J, Danz-Reece M. Ergonomic solutions for the process industries, 1st ed. Burlington, MA: Elsevier; 2004.

[6] Teigen KL. Yerkes-Dodson: a law for all seasons. Theory Psychol 1994;4:525. Available from: http://tap.sagepub.com/content/4/4/525.

[7] Wickens C, Lee J, Liu Y, Becker SG. An introduction to human factors engineering. 2nd ed. Upper Saddle River, New Jersey: Pearson Prentice Hall; 2004.

[8] Parker S, Wall T. Job and work design: organizing work to promote well-being and effectiveness. Thousand Oaks, California: Sage; 1998.

[9] http://www.ors.od.nih.gov/sr/dohs/HealthAndSafety/Ergonomics/Pages/exercises.aspx.

[10] Chemical Safety Board (CSB). Investigation report no. 2004-10-I-IL; 2007. Available from: http://www.csb.gov/assets/1/19/Formosa_IL_Report.pdf.

[11] ASQ website. Available from: http://asq.org/learn-about-quality/process-analysis-tools/overview/mistake-proofing.html.

[12] Chengalur S, Rodgers S, Bernard T. Kodak's ergonomic design for people at work, 2nd ed. John Wiley & Sons; 2004.

Employee training

13.1 INTRODUCTION

A key part of a fatigue risk management system (FRMS) is training employees and supervisors in the biology of sleep and the management of fatigue as a safety hazard. Training is considered a vital part of an FRMS program outlined in ANSI/API RP 755 and is required in the various safety regulations on FRMS. (These regulations are discussed in chapter: Fatigue-Related Regulations and Guidelines). This chapter is intended to provide those responsible for this training with the material, resources, and strategies needed to successfully deliver training on this topic. It may seem odd that people who have slept every day of their lives need training on sleep. Yet many of us do not understand the biology of sleep, the fact that our bodies are "programmed" to seek sleep or resist sleep at different times of the day, and what factors can contribute to our sleepiness. There are misconceptions about the importance of sleep and a lack of awareness of the health consequences of too little sleep. For some individuals, there is even bravado about not needing or not getting sufficient sleep.

Employees who are covered by an FRMS should receive training in sleep and associated policies. This would likely include individuals who work in extended shifts, rotating shifts, and/or might be called to work during their off duty hours. Managers responsible for scheduling and supervising these workers also need FRMS training. The training can be structured as two separate trainings or both groups could attend the same training and managers then receive additional training beyond that given to the employees. The training should be documented by the organization's human resources or safety department. These records would likely be reviewed when the FRMS is audited or an incident potentially involving fatigue is investigated. The training should be repeated periodically, such as yearly. The follow-up training could be a repetition of the original training or shorter refresher training.

13.2 ADDRESSING FRMS TRAINING RELUCTANCE

It is a rare employee who enjoys safety training. Many organizations start training sessions with a statement from management about the importance of the training whether it is safety training or not. This can be done in person by a quick visit and a few words from a member of upper management or in a written letter or other document highlighting the importance of the training the workers are receiving. Those conducting FRMS training may face a higher level of resistance, from both workers and supervisors, than with other training classes. An employee may come to the training with the mindset that an FRMS is trying to dictate what employees do in their time away from work; that the organization has no right to try and control the employee's lives outside the workplace. Trainers need to be prepared to address this line of thinking. The message to give to workers is that FRMS is not about controlling workers' personal time but ensuring workers arrive at work fit for duty. It is setting the expectation that workers are responsible for managing their sleep and reporting potential fatigue-related risks.

Another issue that may come up during FRMS training is why are fatigue and sleep suddenly receiving attention? Or why is this training needed when the organization has been fine without this training for years and years? There are a multitude of reasons for this. First, our scientific understanding of sleep and the risk associated with fatigue has grown in recent years. The medical profession is diagnosing sleep disorders, such as sleep apnea, at a higher rate than ever. Our understanding of the prevalence of sleep disorders and importance of sleep continues to grow with each new study. Scientists have developed a much better understanding about the relationship between sleep and optimal performance in recent years. One key finding is that cognitive deficits continue to accrue due to too little sleep, even though the person may not feel sleepier. Science has shown there are health implications related to sleep deprivation including cardiovascular disease, diabetes, and potentially Alzheimer's; all of which can impact the worker's health and the company's health insurance costs. Second, many people are getting less sleep than in the past. More and more people are working in unusual shifts and longer hours, making fatigue a growing problem. Life in general seems to have gotten busier for many. Third, government regulations and industry guidelines are moving toward a greater understanding of fatigue-related risks and the importance of FRMS. Workplace cultures and attitudes toward sleep are changing. It is slow, but overall people tend to be taking their sleep more seriously. Finally, we have some solutions for sleep-related problems, even if they are somewhat rudimentary solutions. Management can improve the situation by implementing policies and improving work schedules. Workers can improve their sleep by taking some simple steps to improve their sleep environments and bedtime routines and/or seeking medical solutions.

Individuals attending safety training on a chemical process are not going to chime in and say they have been operating a chemistry lab their entire life and know the science better than the instructor. Yet people who believe they know about sleep and how the lack of it affects them are common. Workers may claim they can "get by on fewer hours of sleep" or use "tricks" such as listening to the music or rolling down a car window while they drive to keep themselves alert. Such thinking is counter to the science of sleep. Fatigue impairs performance even if we cannot detect it and people with a significant sleep debt can fall asleep when they are not expecting it. One strategy can be to have someone diagnosed with sleep apnea discuss how much better they did once they began using a CPAP. These individuals typically did not realize how fatigued they were until they were able to get a restful night sleep with the help of the medical device. Examples of these stories are available at http://www.sleepapnea.org/learn/personal-stories.html. Another strategy is to have a worker share a near-miss incident, such as falling asleep while driving home from work.

13.3 TRAINING TOPICS

The FRMS training should be tailored to the specific organization, however, there are some common topics that should be included. These topics and the associated learning objectives are listed in Table 13.1.

Learning objectives are a key part of developing any training. They assist in determining what to include in the training materials. The objectives are also key in developing tests to assess how well the students learned the material after the training.

13.4 WAYS TO ENGAGE TRAINEES

Everyone has a limit to their attention span during a lecture. Discussion questions can be useful to change the format of presenting material during training. Questions can be used for a large group discussion among class members. The same questions can be used in small group discussions or self-reflection. Allowing students to list examples or share with a few people before sharing with the entire class can be less threating for some and increase participation in a follow-up discussion. Another benefit of discussion questions is that it can help students to see applications of the training material. When a class member shares his or her experiences, a sleep disorder or a dangerous situation caused by being overly fatigued, students may begin to accept the

Table 13.1 Common FRMS Training Topics

Topics/Subtopics	Learning Objectives (To Be Able To...)
FRMS policy ■ Motivation for the policy (improved safety, industry standard, government regulations, related accidents and incidents, etc.) ■ How the policy fits into the organization including who in the organization is responsible for its development and implementation and how it integrates with other policies and procedures within the organization	■ Explain how fatigue is related to safety ■ Reference the organization's FRMS ■ Describe how the FRMS fits into the organization's overall safety program
Biology of sleep ■ Importance of sleep ■ An overview of sleep debt, circadian rhythms, sleep stages, and the need for sleep ■ Common sleep disorders ■ Effects of fatigue on health and human performance ■ Effects of fatigue on quality of life	■ List the important functions that occur during sleep ■ Describe physical processes involved in sleep and fatigue ■ Describe common sleep disorders ■ Explain how time of day, time on task, and time since sleeping affect human performance and sleepiness
Sleep strategies ■ An overview of sleep hygiene factors that can improve sleep ■ An overview of bedtime routines that can make falling asleep easier	■ List tips for achieving good quality sleep ■ List strategies that can improve the process of falling asleep
Fatigue risks ■ Signs of fatigue ■ Accident potential ■ Countermeasures	■ Identify fatigue in yourself and others ■ Understand the serious of fatigue risks
Work schedules ■ Hours of service ■ Work shifts ■ Unplanned work	■ State the organization's hours of service rules ■ Identify challenges associated with evening and night shifts, long hours, and rotating shifts ■ State the organization's policy on working emergency or call-out duties and the related time off for these

importance of the training and have a greater appreciation of how the ideas presented can be applied. The best discussion may happen spontaneously, but trainers may find it useful to have a number of questions prepared to use when students seem uninterested or not engaged in the material. The following are discussion questions that could be used through FRMS training.

FRMS discussion questions:

■ What are some challenges that you or your coworkers have experienced as a result of shift work or working for long hours?

■ What are some symptoms of fatigue you have experienced including physical, mental, and emotional symptoms?

- What are some hazards and potential consequences if someone attempted to perform a job in your unit while overly fatigued?
- Can you recall a time you suffered from sleep deprivation (ie, "pulling an all-nighter," caring for a sick infant)? How well were you able to function?
- Does the amount of sleep or when you sleep vary with your work schedule? How might that affect you?
- What are positive aspects of your bedroom that help you sleep? What improvements could you make to improve your sleep?
- Do you commonly take naps? How does that affect your ability to get enough sleep and balance other parts of your life?
- What are your greatest challenges to getting a good night's sleep? Have you overcome barriers to sleep in the past?

13.5 TRAINING FOR SUPERVISORS

While the workers' training will focus on the biology of sleep, the effects of fatigue, and the FRMS policy; the supervisors' training should have an additional focus on the application of the FRMS and scheduling workers. Supervisors will typically be responsible for the unit's compliance with the FRMS. They will need to obey the hours of service limits in the policy, and this may affect staffing levels. It could also affect how maintenance and other unusual tasks are scheduled. More closely managing work hours to ensure everyone has sufficient time away from work for proper rest may affect how vacations, training, and other nonwork hours are scheduled, as well. Any other considerations involved in work schedules such as union agreement, seniority preferences, and staffing requirements should also be addressed in the training. As a new FRMS is implemented, supervisors may have many questions about how existing scheduling practices will change under the new policy.

Beyond just the compliance aspects of the new FRMS, supervisors may also find themselves in a sales role to convince the workers that fatigue risk management is an important issue that must be managed. As with most safety training, the workers judge the seriousness of the training from the supervisor's buy-in after the training and their peers' response to the training. If supervisors believe in the importance of the FRMS and echo that to their workers, then the implementation of the safety policy will be much more effective. This may mean more time is needed to justify the need for the FRMS in the supervisor training. It may also be beneficial to train the supervisors before the workers.

For many supervisors, the hours of service aspects of the FRMS may be the most familiar and the easiest to implement. The need to observe workers for signs of fatigue and be prepared to address these will be a new duty for many supervisors. In some situations, workers may be reluctant to admit to being fatigued to avoid being considered unable to handle the workload by their co-workers. Taking time off to rest can have a negative effect on the workers' take home pay and that can also make workers hesitant to reduce their work hours. At times, supervisors may have the opposite concern that workers might be inclined to abuse the FRMS policies to take extra time off from work. Events such as the local sports team being televised, the beginning of hunting season, holidays, or other times that it is desirable for people to be off work might tempt some workers to claim excessive fatigue. Trainers should be prepared to address these potential concerns with supervisors receiving FRMS training.

13.6 FREELY AVAILABLE FRMS TRAINING MATERIALS

The National Institute for Occupational Safety and Health (NIOSH) has online training available for nurses in the area of FRMS [1]. The training provides information about the risks of shift work, long work hours, and strategies to reduce these risks. The 12 h-training is divided into two parts and individuals can receive a certificate of completion.

The training has the following learning objectives.

1. Explain why scientists think shift work and long work hours are linked to health and safety risks.

Table 13.2 Example Case Study Discussion

Situation	Discussion Questions	Responses
A production facility operates 24 h a day, with three shifts, morning, afternoon, and evening.	■ How realistic is this scenario?	■ Will vary by organization
All shifts are permanently allocated to three sets of workers.	■ What are the advantages and disadvantages of this arrangement?	■ Workers can maintain a set wake/sleep schedule ■ Harder to staff evening shift
There is no limit placed on the number of consecutive nights operators could work and there are fewer staff scheduled to work at night than in the day.	■ What are the potential problems with this?	■ Not enough recovery time for workers ■ More demanding work on night shift
A review of injuries, near misses and incidents revealed high numbers of problems on the night shift.	■ What are the potential causes of this? ■ What changes should be made?	■ Inexperienced workers, less supervision, fatigue ■ Implement at FRMS system, limit consecutive work days, review staffing levels

2. Identify the health and safety risks that are linked to shift work and long work hours.
3. Identify individual factors that can lead to differences in a nurse's ability to adjust to shift work and long work hours.
4. Discuss management strategies to improve the design of work schedules and to improve other aspects of the organization of the work.
5. Discuss strategies that nurses can use in their personal lives to reduce the health and safety risks.

The computer-based training is geared to nursing; however, some aspects of the material could be useful when preparing FRMS training for another industry. True/false quizzes are included throughout the training which could be a useful example of assessment questions to be used in the training process.

Safe Work Australia [2] has case studies available in their online publication *Guide for Managing the Risk of Fatigue at Work*. Table 13.2 is a

---

EXAMPLE: SLEEP DEPRIVATION EXERCISE

The exercise should be introduced with some definitions:

- *Hours of wakefulness* is the total number of hours you have been awake.
- *Sleep debt* is the accounting of accumulated lost sleep that must be repaid by sleeping to achieve full alertness and wakefulness.
- *Work day* is the sum of hours of wakefulness and sleep debt.

A person's mental performance level can be compared for work day to alcohol consumption using the following table [3]. Drivers are considered legally drunk in many places when their BAC is 0.08.

Work Day (h)	Equivalent Blood Alcohol Concentration
10	0.00
12	0.01
14	0.03
16	0.05
18	0.07
20	0.09
22	0.10
24	0.14

For example, if you are on day 4 of a newly changed work shift and have gotten 3 h less sleep per night for the last four nights, your sleep debt is 12 h (4 days $\times$ 3 h/day = 12 h). For 8 h in your work shift, you would have a 20-h work day (12 h sleep debt + 8 h of work = 20 h work day). From the table, your mental performance is equivalent to a legally drunk driver with a BAC of 0.09.

modified version of one of their case studies that could be used to encourage discussion during FRMS training.

The website Wilderness.net is a collaboration between the University of Montana, the Arthur Carhart National Wilderness Training Center, and the Aldo Leopold Wilderness Research Institute. Among their online training materials, they have a tool that compares a person's mental performance after varying hours of wakefulness to an equivalent blood alcohol concentration (BAC) [3]. Although there is always variability among humans, this analysis can be a useful training exercise to emphasize the dangers of sleep deprivation.

EXAMPLE: FATIGUE RISK EXERCISE IN TRAINING

Your organization is eager to research a production process after a maintenance shutdown that took longer than scheduled. Your regular shift is Monday through Friday 8:00 to 5:00. You typically get 8 h of sleep at night (11:00 pm–7:00 am). You have been approved to work 3 extra hours of overtime each day. Because of this overtime, your commute, and family obligations you get two less hours of sleep each night and are getting to bed at 1:00 am. What is your workday and equivalent BAC at the end of each day?

Solution

Monday at 7:00 pm your hours of wakefulness is 12 h (from 7:00 am to 7:00 pm). Since you got your regular night sleep on Sunday night, your sleep debt is 0 h. This results in a workday of 12 h and an equivalent BAC of 0.01.

Tuesday at 7:00 pm your hours of wakefulness is 12 h (from 7:00 am to 7:00 pm). Your sleep debt is 2 h since you went to bed later on Monday night. This results in a workday of 14 h and an equivalent BAC of 0.03.

Wednesday at 7:00 pm your hours of wakefulness is 12 hs (from 7:00 am to 7:00 pm). Your sleep debt is 4 h since you went to bed later on Monday and Tuesday nights. This results in a workday of 16 h and an equivalent BAC of 0.05.

Following this pattern, your workdays are 18 h on Thursday and 20 h on Friday with an equivalent BAC of 0.07 and 0.09, respectively.

The take away from this exercise for students should be that the accumulation of sleep debt can have a significant effect. Consistent failure to get a full night's sleep can greatly increase the fatigue-related risks. It is also worth remembering that sleep debt recovery will typically not occur after a single sleep period.

13.7 **FRMS TRAINING ASSESSMENT**

Key to any safety-training program is assessing how well the students have met the learning objectives. Quizzes are often used to accomplish this. The following are sample questions and feedback tied to the learning objectives stated in Table 13.1.

Question 1: What are the benefits of an FRMS in the workplace?

a. Improved safety
b. More effective workers
c. Fewer errors
d. Improved worker well-being
e. All of the above

This question evaluates the learning objective: Describe how the FRMS fits into the organization's overall safety program

The correct answer is 5 (all of the above). An effective FRMS will reduce fatigue hazards, making the workplace safer. Workers who are getting sufficient sleep tend to be more effective, make fewer errors, and have better overall health and quality of life.

Question 2: Which of the following is the most significant concern for a worker with sleep apnea?

a. Frequent, loud snoring
b. Difficulty falling asleep
c. Getting less restorative sleep
d. Low blood pressure
e. Chronic lung damage

This question evaluates the learning objective: Describe common sleep disorders.

The correct answer is 3 (getting less restorative sleep). Snoring is a common symptom of sleep apnea. Difficulty falling asleep is associated with insomnia. High blood pressure and other heart-related problems can occur with sleep apnea, not low blood pressure. Although sleep apnea causes someone to wake up due to breathing difficulty, it does not damage the lungs.

Question 3: Which of the following is a recommend strategy to help shift workers sleep during daytime hours?

a. Have your bedroom dark
b. Have your bedroom warm
c. Watch television to relax before sleeping
d. Have an alcoholic drink
e. Turn off any source of steady sound such as exhaust fans

This question evaluates the learning objective: List strategies that can improve the process of following asleep.

The correct answer is 1 (have your bedroom dark). Light stimulates wakefulness both directly and indirectly through the actions of circadian rhythms. A cooler bedroom also helps you sleep. The blue light from televisions and computer screens can have the opposite effect on your body's circadian rhythms; the focused attention on the television and the noise may also distract someone from going to sleep. Experts do not recommend alcohol as a sleep strategy; it can have a disruptive effective on sleep patterns and the quality of sleep. Steady, unobtrusive sound (white noise) can help mask unwanted sounds that can keep someone wake or woke someone who is asleep.

Question 4: Which of the following is not a common symptom of fatigue?

a. Involuntarily nodding off
b. Yawning
c. Lowered blood pressure
d. Poor hand–eye coordination
e. Trouble focusing

This question evaluates the learning objective: Identify fatigue in yourself and others.

The correct answer is 3 (lowered blood pressure). Sleep deprivation is associated with elevated blood pressure. The other answers are common fatigue symptoms.

Question 5: Difficulty falling asleep because the body's circadian rhythms are not synchronized is more common with which of the following?

a. Day shift work
b. Evening shift work
c. Night shift work
d. Rotating shift work
e. None of the above, your body is automatically able to synchronize circadian rhythms

This question evaluates the learning objective: Identify challenges associated with evening and night shifts, long hours, and rotating shifts.

The correct answer is 4 (rotating shift work). The body's circadian rhythms can get out of sync. Having a consistent work and sleep schedule (even one opposite the standard working during the day and sleeping at night) can allow the body's natural sleep drive to match the work schedule, although

I recognize and pledge my commitment to work safely and come to work well rested every day.

I acknowledge that no task is so important that it will be completed without regard to safety and the well-being of others.

I am the person most aware of my actions and best able to evaluate whether I am fit and alert to work. I have a responsible to inform my supervisor if anyone is making poor decisions, failing to follow best safety practices, or too tired to work, including myself.

Signature: _____

Date: _____

■ FIGURE 13.1 Sample safety commitment.

some workers never adjust these schedules. This is much more difficult when the work schedule is frequently changing such as is the case for a rotating work shift.

A major goal of any safety training is to change employee behavior. Having employees sign a safety pledge or another form of commitment to change their behavior can be an important part of an FRMS safety training. A sample pledge is provided in Fig. 13.1.

Another potential source of assessment material is from Transport Canada [4]. Their online FRMS training material includes a list of training review questions that could be useful when preparing an FRMS training. A modified list from their training material includes the following:

- Name three aspects of your life that can be affected by nontraditional hours of work.
- Name two types of biological rhythms that are regulated by the body clock.
- What are two major causes of fatigue?
- Name four symptoms of fatigue.
- Compare performance in the following situations: (1) being awake for over 17–23 h (2) being under the influence of alcohol
- On average how many hours of sleep should you ideally get each night?
- How can "white" noise positively affect sleep?
- What room temperature range is most suitable for promoting sleep?
- Explain the positive effect of a prebedtime routine.
- Explain what sleep inertia is.
- Name two benefits of napping.

- Why does caffeine lose its ability to improve your alertness if you drink caffeinated drinks regularly?
- Name three tips you can apply to get the maximum benefit from caffeine as a stimulant.
- Name three problems associated with sleeping pills.
- Name four types of health complaints often reported by shift workers.
- What effect can physical exercise have on sleep?
- Describe two strategies that can help in balancing work and family.
- Name three factors you should consider when designing work schedules.
- Name two employee responsibilities and two employer responsibilities with respect to managing fatigue-related risks.
- Name one negative aspect for each of the following shifts: morning, afternoon, and night.

REFERENCES

[1] NIOSH, Caruso CC, Geiger-Brown J, Takahashi M, Trinkoff A, Nakata A. NIOSH training for nurses on shift work and long work hours. (DHHS (NIOSH) Publication No. 2015-115). Cincinnati, OH: US Department of Health and Human Services, Centers for Disease Control and Prevention, National Institute for Occupational Safety and Health; 2015. http://www.cdc.gov/niosh/docs/2015-115/

[2] Safe Work Australia. Guide for managing the risk of fatigue at work 2013. http://www.safeworkaustralia.gov.au/sites/SWA/about/Publications/Documents/825/Managing-the-risk-of-fatigue.pdf

[3] https://www.wilderness.net/toolboxes/documents/safety/Fatigue%20Case%20Study.pdf

[4] Transport Canada Training. Fatigue management strategies for employees, TP 14573E; 2007. https://www.tc.gc.ca/Publications/en/tp14573/pdf/hr/tp14573e.pdf

Chapter 14

Naps

Throughout this book, there have been descriptions of what happens when a person is sleepy because they missed a night's sleep or because they repeatedly do not obtain adequate sleep. People can miss out on sleep from a variety of sources, including their choices to stay awake or go to sleep, external forces, such as a newborn, an individual's genetics, or sleep disorders. In an industrial setting, shift work or long work schedules are a likely source of fragmented and inadequate sleep duration. The resulting elevated sleepiness can cause increased mistakes, worker accidents and injuries. Numerous countermeasures may help an employee reduce or temporarily mask the effects of sleepiness and allow an employee to perform as if they were not sleep deprived. Since we do not know the restorative molecular events that occur with sleep, there is not a pill or remedy that can mimic the effects of sleep. Therefore, sleeping is still the only way to obtain the full benefits of sleep.

14.1 PERCEPTIONS OF NAPPING

Traditionally, the goal has been to improve duration and consolidation in the primary sleep period. An alternate strategy to improve sleep durations and the benefits of sleep is to add naps into the overall sleep strategy. Napping, even very short naps, have been shown to improve cognitive performance and reduce sleepiness. In the past, the concept of napping has been stigmatized as the notion of napping has been equated with laziness. This view is becoming more and more antiquated as researchers gather increasing amounts of data that demonstrate the benefits of napping. Napping may fill a vital role as a countermeasure against sleepiness and the accompanying performance risks. Effective napping can reduce the burden from both inadequate sleep from sleep deprivation as well as from some sleep disorders. Given all of the things that impinge on our nighttime consolidated sleep, it may be wise to design strategies to incorporate strategic napping into our daily routine.

Napping has been growing as an acceptable part of corporate culture. Scientists have been continually identifying new benefits to napping and improving our understanding about how to strategically employ napping in the workplace. Improvements in cognitive performance, productivity, motivation, and emotional composition have thus far been observed with napping. Famous nappers include politicians and businessmen, such as Winston Churchill, John F. Kennedy, Napolean Bonaparte, John D. Rockefeller, and Lyndon Johnson. Often these people scheduled daily naps to reduce their sleep debt that built up in the first half of the day and improve their cognitive and emotional well-being during the second half of long work days. Creative-type people also employed naps as part of their daytime activities. Leonardo DiVinci, Thomas Edison, Salvador Dali, and Albert Einstein were known to devote time to napping when needed due to previous periods of sleep deprivation or because of irregular sleep patterns that made napping ideal. Although napping may not ensure success, it shows that napping does not necessarily interfere with being successful in one's chosen field.

Given the importance and benefits of sleep, a napping philosophy is slowly making its way into some corners of corporate culture. As work schedules become more erratic and the production schedule continues on its 24-h cycle, the recognition that naps can increase performance has become more accepted. Businesses are being set up to plan for and include areas for employees to refresh with a nap. A National Sleep Foundation survey of employees found that 34% of workplaces allowed them to nap and 16% provided a place to nap, such as either specific nap rooms or designated nap areas [1,2]. Some of these institutions include the following:

- AOL-Huffington Post
- Ben and Jerry's
- Capital One Labs
- Cisco
- Google
- National Aeronautics and Space Administration (NASA)
- New York Times
- Nike
- Pricewaterhouse Coopers
- Procter & Gamble
- Uber
- Zappos

Clearly, these entities demand a lot from their employees who are dedicated to both the success of the institution as well as their own career success. Taking a break and restoring one's cognitive performance and alertness with a short nap can pay dividends in the long run in improved employee satisfaction, reduced turnover rate, and ultimately increased productivity.

This chapter will cover the benefits of napping, considerations when trying to institute a strategic napping process for an individual or workplace, and the experiences and history of napping. By the end of the chapter, we hope that one is asking if their operation would benefit from napping and if there is an opportunity to institute a napping program within the physical facilities.

14.2 IS SLEEPINESS A PROBLEM AT WORK?

In 2008, the National Sleep Foundation polled 1000 individuals working 30 h or more per week from every region of the country to understand how sleep influences work and how work influences sleep [2]. Respondents represented workers from blue-, gray-, and white-collar groups. The following incidents occurred at least a few nights a week in the last month. Key findings from study include those that follow.

14.2.1 How much sleep and what quality of sleep are workers getting?

- On average, working individuals obtain 6 h and 40 min of sleep on nights prior to work but feel they need 7 h and 18 min, a difference of nearly 40 min each night.
- Nearly 1 in 5 respondents stated they get 2 h less sleep during the week than on weekends. This is associated with both an increased rates of drowsy driving and having sleepiness interfere with daytime activities.
- 65% reported a sleep problem; 44% report this occurring every night or almost every night.
- 49% reported feeling unrefreshed after waking in the morning.
- 42% were awake a lot in the night.
- 29% woke up too early without being able fall back asleep.
- 26% had difficulty falling asleep.
- 12% say their current work schedule does not permit them to get enough sleep. This was true across all types of employees. Longer work hours are associated with increased body mass index (BMI).
- 14% missed at least one event because of sleepiness.
- 20% agree that they have sex less because of sleepiness.
- 8% use alcohol, 7% use over-the-counter medications, 3% use prescription medications, and 2% use herbal supplements as a sleep aids at least a few nights a week.

Therefore, sleep problems are pervasive throughout the workforce and one's work schedule can contribute to the sleep problems. A lack of sleep affects daytime performance and quality of life for these employees.

14.2.2 What are the consequences of the disrupted and inadequate sleep?

- 33% said there was a moderate chance of falling asleep while sitting and reading.
- 28% reported that sleepiness interfered with daily activities with 5% saying it does so every day.
- 36% have driven drowsy at least once per month during the past year.
- 26% had incidence of drowsy driving on their commute to or from work.
- Nearly 11% have fallen asleep while driving, with 2% have an accident or near miss due to drowsiness.

Work-specific consequences

- 40% report becoming impatient with others at least a few days in the last month.
    - Occurred in 29% of people that worked 50 h or more, 20% of people that worked 40–49 h, and 12% of people that worked 30–39 h.
- 27% found it difficult to concentrate with less sleep associated with a more frequent occurrence.
- 20% noticed their productivity was lower than expected at least a few days a month.
- 16% had difficulty organizing work
    - 9% of 50 h work weeks, 4% in 40–49 h work weeks.
    - Twice as often in white collar workers.
- 9% had to do a task over due to mistakes.
- 11% injured themselves or someone else or was in a serious incident or accident
    - These incidents were associated with longer sleep latency, less than 6 h of sleep on workdays, and 1 or more naps per month due to sleepiness.
- 29% fell asleep or became very sleepy at work because of a sleep problem.
- 12% were late to work.
- 4% left work early.
- 2% did not go to work because of sleepiness.

14.2.3 How do people deal with inadequate sleep?

- 80% said they just accept it and keep going
    - Drowsy driving at least once a month over the past year.

❑ 80% of the survey participants say that they drink 2 or more caffeine drinks.

❑ 5% are likely to use alerting medications.

■ 37% will take a nap to combat sleepiness, other sleep adjustments include going to bed early to make up sleep and increasing sleep on the weekend.

14.2.4   Is napping appropriate for the workplace and would individuals actually nap?

■ 10% say they had napped at work.

■ 34% say their employers allowed them to nap at work.

■ 16% said their employers provided a place for employees to nap.

■ 26% of those surveyed said they would take a nap at work, though their employer does not allow it at the current time.

The aforementioned data were designed to illustrate the potential problems that inadequate sleep can have in the workplace. Inadequate sleep is likely more pervasive than people may be aware of. This can manifest as more strained interpersonal relationships, decreased productivity, and an increased propensity for accidents. Based on the data, employees are aware that they may not be performing at their peak throughout the day and that their safety and productivity would benefit from more sleep. The question then becomes philosophy: would it be in the best interests of the organization to include napping as a remedy or not? Based on the experiences of companies listed earlier, napping is a viable countermeasure to the lost productivity with sleepiness, though this may not be the solution for all. Our aim is to present the benefits and to encourage organizations to determine if a napping policy is appropriate for their employees.

14.3   BENEFITS OF NAPS

Given the amount of accumulated sleep debt present throughout the population, napping can provide numerous benefits to the employee and improve the productivity in numerous areas of business from the office area to the factory floor and decrease the risk of accidents and potentially severe incidents. The bulk of the scientific literature supports the fact that naps are beneficial.

1. Decreased sleepiness—Sleepiness is a subjective parameter that is measured using one of many self-rating scales, such as the Stanford sleepiness scale (SSS) or the Epworth sleepiness scale (ESS). Napping should dissipate some of the sleep debt, which will reduce the

propensity to go to sleep. Using these techniques it has been shown that sleepiness is reduced under a variety of conditions:

    a. Naptimes from 10 min to over 90 min have been shown to reduce sleepiness. Longer naps will reduce sleepiness for a longer period, but the longer the nap may result in sleep inertia that will make the person feel groggy.

2. Increased alertness and vigilance—This characteristic is often measured using the psychomotor vigilance test, which measures how quickly one responds to a light that flashes at inconsistent intervals. Reaction time and lapses, in which the subject does not respond within 0.5 s, are measured.

    a. After a relatively normal sleep of 7.4 h, a 15-min nap led to brain signatures that indicated increased alertness [3].

    b. Either a 15 or a 45-min nap increased alertness for 3 h compared to no nap [4].

3. Increased cognitive performance—There are numerous ways to test cognitive performance, but many of the methods use some type of sequential math problems or substitution task.

    a. After only 4 h of sleep night, a 15-min nap opportunity improved logical reasoning [5].

    b. A 10-min nap increased cognitive performance. Surprisingly, a 10-min nap can be better than 30-min nap on metrics of alertness and performance [6].

4. Better emotional state and increase in patience

    a. After only 4 h of sleep night, a 15-min nap improved results on a depression scale [7].

Napping can also improve the well-being and safety of individuals with sleep disorders. A daytime nap can help dissipate the sleep debt. In the case of narcolepsy, a scheduled nap may help prevent sleep attacks throughout the day. In addition, napping may help with dissipating sleep debt for individuals with sleep apnea and insomnia, if it is possible to nap. Though, one may not be able to schedule a nap under these circumstances.

Napping itself has been shown to be beneficial. Yet, napping too often can be an indication that one is not getting enough restorative sleep or that there may be some other health problem. So while napping has numerous benefits, excessive napping may be an indicator that one needs to see a physician.

14.4 STRATEGIC NAPPING

While napping can be an effective countermeasure to sleepiness, several questions arise. How long does one need to nap to get positive benefits? How much benefit is there to a nap? How long do the benefits of a nap last? In the past decade and a half, sleep researchers have begun to address these questions so that recommendations can be made to help people cope with times they are not permitted to get an adequate amount of sleep. It would be impossible to design a specific course of action that encompasses every situation. Instead, principles presented here can be used to devise the best strategy for the individual. When napping, several factors should be considered including those given in subsequent sections.

14.4.1 Nap duration

A nap does not need to have to be a long nap for the person to experience benefits. Naps as short as 10 min have been shown to have beneficial effects on alertness, vigilance, and cognitive function. Longer naps can have increased benefits, but the onset of benefits may take longer because of sleep inertia [8]. Sleep inertia is the "foggy" feeling one has after just waking up, which can impair cognitive performance (see section later). Naps that last up to 20 min will provide numerous and increasing benefits while reducing the effects of sleep inertia, which cancel out the short-term benefits of the nap. If a longer nap is needed, then timing it with the natural sleep cycle of 90 min should help to increase the likelihood of waking in a lighter stage of sleep and reducing the possibility of sleep inertia (Fig. 14.1).

14.4.2 When the nap occurs during the day

The time of day that the nap is taken can change the ability of a person to get extended sleep or can change the stages of sleep that occur during the nap. In other words, trying to sleep when the circadian output is trying to arouse the person may result in a truncated sleep duration. Because sleep stages may be different depending on the time of day, a nap in the morning may look different from a nap taken late in the afternoon. The naptime that shows the most restoration is between the hours of about 1 pm and 3 pm, when the circadian output is lowest. Interestingly 25-min nap at 12:20

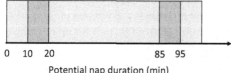

Potential nap duration (min)

■ FIGURE 14.1 Rationale for nap durations. Times in light gray (yellow in web version) have nonideal characteristics whereas time points in dark gray (green in the web version) would work for either short and longer naps. 0–10 min, sleep onset latency may not actually allow people to nap. 10–20 min: ideal for short naps with benefits for sleepiness and performance; 20–85 min, improvements in sleepiness and performance; drawbacks include increased likelihood for sleep inertia; 85–95 min, ~90 min is a good duration for a longer nap given improved cognitive performance and sleepiness. 90 min is about 1 full sleep cycle and the individual can wake up around stage NREM1, minimizing sleep inertia; 95+ min, restorative sleep with an increase likelihood of sleep inertia.

resulted in subjective and objective improvements in sleepiness, but individuals did not see as much cognitive benefit. Yet at 2:00 pm a 25-min nap improved alertness and cognitive improvements [7,9]. Thus, there may be subtle differences in cognitive benefit depending on the timing of the nap. Future research will help define differences in these characteristics.

The cautionary note about napping at 2:00 pm is that it may be more likely to result in sleep inertia upon waking. Shiftworkers may be more likely to nap during the other circadian low point from 3 to 6 am, when it is easiest to sleep because there is little circadian opposition. Though naps taken during this period may be subject to considerable amounts of sleep inertia [10]. Thus, the timing of a nap should fit with both the goals to reduce sleep debt in the most efficient manner and permit the person to fall asleep for the desired amount of time without adverse effects on the rest of the daytime schedule.

14.4.3 Future sleep debt

Taking a nap can dissipate sleep debt at times outside of the normal sleep period. Thus, planning a nap should attempt to leave enough time for sleep debt to accrue so that the person may sleep again at a normal bedtime. Without sufficient sleep debt, the latency to sleep onset may be extended and delay the ability for one to sleep at night. This can lead to frustration regarding sleep and the delayed sleep onset can result in insufficient sleep when one has to wake up the following morning. Thus, the nap can actually cause insufficient sleep if the timing of the nap is not thought out.

14.4.4 Sleep inertia

Sleep inertia is the cognitive decrements that occur when coming out of sleep. Sleep inertia is more likely to occur when awakening from a deeper stage of sleep [11]. Thus, a longer nap, such as one that is more than about 20 min may result in severe sleep inertia. Also, sleep deprived individuals are also at risk for sleep inertia because they will move to deeper stages of sleep faster, which will increase the likelihood of sleep inertia. Sleep inertia can cause decreased cognitive function for up to an hour or more. The decrease in cognitive function from sleep inertia can be more serious than the cognitive decreases seen with sleep deprivation. Even when a person feels back to normal, they may not be at their full cognitive capacity [12]. After a relatively normal sleep of 7.4 h, a 45-min nap led to brain signatures that indicated no change in alertness [3], suggesting that sleep inertia may have set in. Therefore, the person should be aware of sleep inertia when planning for a nap. If the person needs to awaken and be ready to go, the nap should be less than 20 min. But if the person can endure some amount of cognitive grogginess before needing to function, then a longer nap may be a better plan.

These considerations are summarized in Figs. 14.1 and 14.2 for a normal schedule and scenario. It factors in both the build-up of sleep debt and oscillations of the circadian system that promote arousal. When to nap will depend on the individual differences, needs, and circumstance of the person. The considerations for a short nap or long nap are articulated in the legend that accompanies the figures box. These generalizations should be used to balance the numerous variables that can lead to a productive and satisfying nap without too much disruption in the overall schedule (Fig. 14.2).

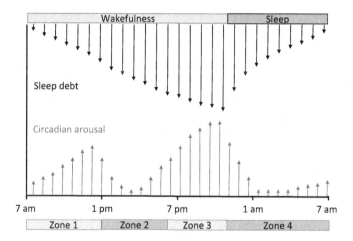

■ **FIGURE 14.2 Classic arrow diagram to design nap strategies under a routine schedule.** Descending *black arrows* indicate the state of sleep debt. *Longer arrows* suggest more sleep debt. *Rising arrows* estimate the magnitude of the arousal signal over a typical day. In this figure, we have broken the day into zones for illustrative purposes. Green indicates easier time to initiate a nap and yellow has some caveats about initiating a nap. An individual's circumstances may be very different and napping may be ideal at a time that is not recommended for a typical schedule, such as for a shift worker. Zone 1, difficult to nap because sleep debt is low due to sleep the night before and circadian factors are increasingly signaling wakefulness. Zone 2: best napping time based on increased sleep debt build up and a reduction of circadian arousal signals. The precise naptime may promote a longer nap because of reduced arousal signals and achieving a deeper stage of sleep leading to sleep inertia. Zone 3: napping during this period may be less than ideal. Sleep debt is increasing, but circadian arousal factors are increasing during the late afternoon and evening. This will promote wakefulness and make it difficult to sleep. In addition, late afternoon or evening napping may dissipate sleep debt to the point that it causes increased sleep onset latency for the nighttime consolidated sleep period. Zone 4: consolidated sleep period and napping during this period would promote longer nap duration that would increase the likelihood of sleep inertia.

14.5 NAPPING RECOMMENDATIONS FOR THE WORKPLACE AND FOR SHIFTWORK

For healthy individuals on a normal schedule, there are several recommendations based on empirical data and those predicted from sleep research that has been conducted.

1. Naptime of about 20 min for a short nap. This duration has been shown to improve performance on word recollection tasks and pilot performance on flight simulator tasks (NASA technical memo 108839). The pilots' sleep times were under 15 min, and the individuals did not report an improvement in their subjective sleepiness. Despite the lack of reduced perceived sleepiness, the pilots exhibited cognitive improvement. Longer naps can be taken, but they are subject to sleep inertia. If a nap needs to be greater than 20 min or so, then try to make it about 90 min. Since the natural sleep cycle is about this long, it is more likely that a person will wake in a lighter stage of sleep, such as non-REM stage 1 sleep. Waking during this point will reduce sleep inertia.

2. Avoid napping at the daytime circadian trough. One may want to avoid napping at the low point of the circadian cycle. Napping at this point may increase the likelihood that one will sleep longer than the 20 min and will be more likely to enter deeper stages of sleep. This will result in sleep inertia when one wakes.

3. Avoid napping too late in the day. Long naps and naps late in the afternoon or evening may delay bedtime and increase the time to falling asleep.

The same aforementioned rules do not necessarily apply to a person on a shift work schedule. Napping can improve the amount of slow wave sleep in shift workers [13]. There are more considerations with the influence of the circadian clock and influences of light, social obligations, and work obligations that may interfere with the ability to nap and managing the sleep schedule (Fig. 14.3).

For people on a second shift:

1. It has been suggested that employees on a second shift are able to get the most sleep because of the schedule. Especially for individuals that are more evening type people, this shift tends to allow the most sleep as the work time matches with their natural circadian rhythms. If a person is on a rotating shift, ideally the person will take advantage of this shift timing and try to obtain the most sleep on this shift while they are assigned to it.

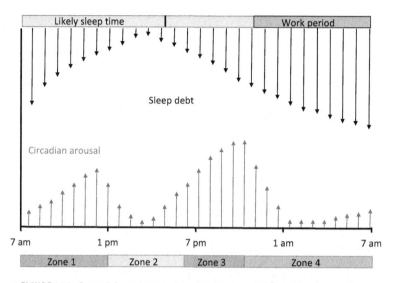

■ **FIGURE 14.3 Potential naptimes and considerations are different for night shift workers.** Sleep debt builds during wakefulness, including throughout the shift. In this example, the circadian rhythms of the worker have not adjusted to the new schedule and continue on a typical schedule. Without a nap, the end of the shift may have the largest sleep debt with little counteracting arousal from the circadian rhythms. Dark gray (green in the web version) indicates a better naptime and light gray (yellow in the web version) has some caveats about initiating a nap. Napping strategies throughout the day are considered as follows: Zone 1: this is the most likely time for a night shift worker to get consolidated sleep to dissipate sleep debt from the previous night's shift. Often sleep during this time is shorter and less consolidated than sleep over night. Zone 2: a less appropriate time to nap under these circumstances. Either sleep debt has been dissipated by consolidated sleep earlier in the day (Zone 1) or a nap initiated during this period may be too far from the start of the shift and significant sleep debt can build up prior to the start of the shift. Zone 3: during this period, it would be recommended to try and nap prior to the start of the shift to dissipate as much sleep debt as possible prior to the start of the shift to maintain the best performance. Zone 4: scheduled work shift. Napping during this time may be beneficial if allowed by the employer. Only a short 10–20-min nap should be taken to improve performance, reduce sleepiness, and avoid sleep inertia.

For people on the third shift:

1. Use strategic napping to dissipate sleep debt just prior to the start of the shift. Given that the third shift occurs primarily at a time when the circadian system is not relaying wake signals, dissipating sleep debt just before starting the third shift, while avoiding too much sleep inertia, will decrease the likelihood of a sleepiness related accident.
2. Try to nap during the shift. Attempting to dissipate sleep debt during the shift with a short nap has been shown to result in a more alert employee. As with daytime naps, these naps should be limited to less than 20 min to avoid the effects of sleep inertia.

TEXTBOX 14.1 NAPPING IN THE WORKPLACE

The website everydayhealth.com listed the most sleep-friendly companies in America. One of the early adopters of napping at work as NASA, starting in the 1990s employees will allowed to take short power naps. Due to the success of "NASA naps" it has become common for pilots on long international flights to take quick naps as well. The list names Huffington Post-AOL offices in New York the most sleep-friendly location. They have "energy pods" which are reclining rest capsules with beds similar to the tables at massage parlors. The same pods are used by Google. Zappos, the online retailer, also has darkened nap room and encourages napping. One wall of the nap room has a glass wall and is part of the company tour for guests. The website reports that 6% of businesses surveyed had an employee nap room and 34% said their employees were allowed occasional naps at work.

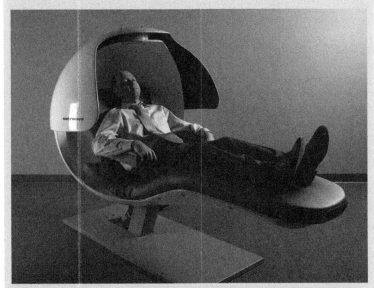

Metronaps Energy Pod i20, reproduced with permission from MetroNaps, www.metronaps.com. All rights reserved.

http://www.podstyle.com.au/metronaps-energy-pod-i20 source

One additional napping strategy that deserves attention was proposed by Horne and colleagues in which one drinks a cup of coffee and then takes a 15-min nap [14]. It takes roughly 15 min for caffeine to take effect. Thus during that time, a person can dissipate sleep debt through a nap and when they awaken, the coffee should have taken effect and continue to be a stimulant to help maintain wakefulness. This strategy may be particularly helpful rather than driving drowsy when only a short nap is possible.

14.6 NAP FACILITIES

14.6.1 Dedicated nap facilities

Some companies have constructed specialized nap pods or dedicated napping facilities. This situation is ideal because employees can use these spaces with confidence to know that they can rejuvenate for the later portion of the work shift (Textbox 14.1). Ideally, these facilities would be an enclosed space that adheres to sleep hygiene recommendations, including providing a dark and quiet space that employees can sleep. These spaces should have an alarm clock to limit naptimes to the desired length and the ability to easily prepare the space for the next person to take advantage and nap.

If it is impossible to set up a dedicated facility but the institution would like to provide some napping accommodation, it would be best to again follow sleep hygiene recommendations. The room should be as dark and quiet as possible. Both external and internal light should be eliminated and the room should be away from loud and irregular noises coming from work or populated portions of the facility. In addition, sleeping quarters should be comfortable and kept up. If it is a shared facility, then there should be a mechanism to apply fresh bedding between occupants

Napping has numerous benefits and can make up for inadequate sleep in the typical consolidated monophasic sleep period that we carry out. Successful napping takes some planning and forethought because haphazard napping may disrupt the regular sleeping schedule. Research is reinforcing the opinion that napping strategies would be beneficial to the employee as well as the organization. These strategies should be seriously considered to determine if they can be integrated into the business environment.

REFERENCES

[1] National Sleep Foundation 2008. Sleep in America poll: summary of findings. Washington, DC; 2008.

[2] Swanson LM, et al. Sleep disorders and work performance: findings from the 2008 National Sleep Foundation Sleep in America poll. J Sleep Res 2011;20(3): 487–94.

[3] Takahashi M, Fukuda H, Arito H. Brief naps during post-lunch rest: effects on alertness, performance, and autonomic balance. Eur J Appl Physiol Occup Physiol 1998;78(2):93–8.

[4] Takahashi M. The role of prescribed napping in sleep medicine. Sleep Med Rev 2003;7(3):227–35.

[5] Takahashi M, Arito H. Maintenance of alertness and performance by a brief nap after lunch under prior sleep deficit. Sleep 2000;23(6):813–9.

[6] Tietzel AJ, Lack LC. The short-term benefits of brief and long naps following nocturnal sleep restriction. Sleep 2001;24(3):293–300.

[7] Hayashi M, Ito S, Hori T. The effects of a 20-min nap at noon on sleepiness, performance and EEG activity. Int J Psychophysiol 1999;32(2):173–80.

[8] Lovato N, Lack L. The effects of napping on cognitive functioning. Prog. Brain Res 2010;185:155–66.

[9] Hayashi M, Watanabe M, Hori T. The effects of a 20 min nap in the mid-afternoon on mood, performance and EEG activity. Clin Neurophysiol 1999;110(2):272–9.

[10] Naitoh P. Circadian cycles and restorative power of naps. In: Johnson LC, Tepas DI, Colquhoun WP, Colligan MJ, editors. Biological rhythms, sleep and shift work. New York: Spectrum; 1981. p. 553–80.

[11] Bruck D, Pisani DL. The effects of sleep inertia on decision-making performance. J Sleep Res 1999;8(2):95–103.

[12] Jewett ME, et al. Time course of sleep inertia dissipation in human performance and alertness. J Sleep Res 1999;8(1):1–8.

[13] Richardson GS, et al. Objective assessment of sleep and alertness in medical house staff and the impact of protected time for sleep. Sleep 1996;19(9):718–26.

[14] Reyner LA, Horne JA. Suppression of sleepiness in drivers: combination of caffeine with a short nap. Psychophysiology 1997;34(6):721–5.

15

Compounds that alter sleep and wakefulness

As humans, we ingest many different compounds that maintain our health and well-being. We consume food to provide energy for our bodies and compounds we believe will help us live longer or improve our health. People take vitamins, unregulated nutritional supplements that modulate health or alertness levels, over-the-counter medications to self-treat small maladies, and prescription medications for serious or chronic ailments. These nutrients and substances may impact sleep either positively or negatively. For example, a decongestant taken for a stuffed-up nose may result in sleep disruption because these processes are regulated by a similar biochemistry but the effects take place in different parts of the body. This is one example of an unrealized side effect that impacts their sleep and wake patterns. On the other hand, people will drink caffeine to purposefully make themselves more alert.

In this chapter, we present some of the commonly used pharmacologics and substances that impact sleep. They may be taken to specifically address sleep problems or indirectly regulate sleep through side effects. We aim to provide readers with a broad introduction and general information regarding some commonly taken compounds. It is in no way meant to substitute for the advice and supervision of a qualified sleep professional or clinician but rather is meant to give an idea of how many compounds can impact sleep regulation. As one integrates new treatments or lifestyle changes, they should be cognizant of how these compounds might impact or alter sleep. By using the information in this chapter, one may become better aware of the impact that these pharmacologic compounds can have on sleep and how one might alter their schedules to obtain the desired benefit of the drug while minimizing the impact on sleep and wake regulation.

15.1 OVER-THE-COUNTER SUBSTANCES

This portion of the chapter describes compounds that are accessible to people to self-administer that might alter sleep and wakefulness.

15.1.1 **Coffee/caffeinated drinks**

Caffeine is one of the most consumed drugs in the world. Over $13 billion is spent each year on coffee alone. Include tea, soda, energy drinks, and all the other sources of caffeine to these figures, and caffeine delivery is a gigantic industry. The desire for caffeine is so large that the food manufacturers have been trying to add caffeine to water and now caffeine gum that can deliver this drug in new and faster ways.

Caffeine is one of the most common countermeasures to sleepiness in today's workforce. It is likely that coffee is available in the office or break room during the mornings and afternoons. Caffeinated beverages are a socially acceptable and extremely effective way to stimulate alertness and focus. One benefit is that it can be titrated to the level that one needs for the particular moment in time. But coffee is not the only way to consume caffeine. There are caffeinated teas, which have roughly half the caffeine of a cup of coffee, and sodas, which can have about a third to a half of the amount of caffeine as coffee (Table 15.1) that can delay sleep and increase alertness [1,2].

15.1.1.1 *Chart of caffeinated beverages*

Of course the actual amount of caffeine consumed depends on the number of servings consumed. For younger people, there is an ever increasing popularity of energy drinks. These drinks can provide a lot of caffeine in an easy-to-drink form, which can result in excessive caffeine consumption. In addition, these beverages contain other substances, such as guarana, a seed with large amounts of caffeine, and taurine, a less studied neurotransmitter in humans, but which may have stimulatory effects on its own [3]. These energy drinks have been the source of controversy because these energy drinks have caused a spike in emergency room visits for elevated heart rate and anxiety in younger individuals [4]. This demonstrates the stimulating power of these particular drinks and how they might interfere with the ability to fall asleep.

Cells, and especially neurons, communicate with one another by releasing specific chemicals (Fig. 15.1). Just as importantly, there must be receptors on the cells that respond to the chemical messages. Imagine a phone call. For information to be communicated, there must be a person talking who initiates a message and a receiver at the other end of the line who listens and comprehends the message. Without both people, there is no communication. In biochemistry, the release of the chemicals is the message sent by the caller and the receptors are the persons on the other end listening and interpreting that message. Some evidence suggests that when certain neurons are working throughout the day in our creative, social, and learning endeavors, they release a compound called adenosine. Adenosine binds to its receptor

Table 15.1 Caffeine Levels in Common Consumables

	Size	Caffeinated (mg Caffeine)	Decaffeinated (mg Caffeine)
Coffees			
Brewed	8 oz (237 mL)	95–200 mg	2–12 mg
Brewed, single-serve variety	8 oz (237 mL)	75–150 mg	2–4 mg
Espresso	1 oz (30 mL)	47–75 mg	0–15
Instant	8 oz (237 mL)	27–173	2–12
Latte or Mocha	8 oz (237 mL)	63–175	
Teas			
Black tea	8 oz (237 mL)	14–70	0–12
Instant iced tea, prepared with water	8 oz (237 mL)	11–47	
Ready-to-drink iced tea, bottled	8 oz (237 mL)	5–40	
Other drinks			
Soft drink	12 oz (355 mL)	0–55	Check the label
Amp energy drink	8 oz (237 mL)	71–74	
5-h energy drink		200–207	
Full throttle energy drink	8 oz (237 mL)	70–100	
Red Bull energy drink	8 oz (237 mL)	75–80	
Rock Star energy drink	8 oz (237 mL)	79–80	
Monster energy drink	8 oz (237 mL)	160	
Caffeinated water	16.9 oz (499 mL)	50–125	
Other products			
Caffeinated gum	1 piece	10–100	
Chocolate chips	1 cup	104	
Chocolate covered coffee beans	28 pieces	336	
Energy mints	2 mints	95–200	
Medications			
Excedrine extra strength	1 tablet	65	
NoDoz max strength	1 tablet	200	

and causes the receptive neurons to reduce their activity. This reduced activity is hypothesized to underlie increased sleepiness [5–7]. Caffeine blocks adenosine from binding to the adenosine receptors [8]. Thus, the message that slows the firing of the receiving neurons is not received and the person remains more alert. Though the person feels more alert, sleep debt continues to build during this period and one may feel even more sleepy when the caffeine wears off [2].

While caffeine is often necessary because many of us are sleep deprived to begin with, the use of caffeine can put an individual into a positive feedback

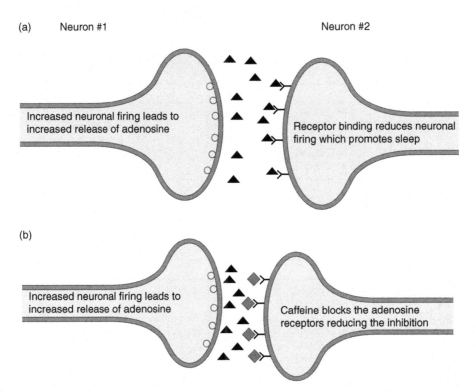

■ FIGURE 15.1 Caffeine is an adenosine receptor inhibitor to promote wakefulness. (a) Under normal circumstances, Neuron #1 will fire throughout the day due to waking activities, which generates adenosine [black (red in the web version) triangles] in the synaptic cleft between the two neurons. The adenosine binds to a receptor on Neuron #2, which reduces the firing of that neuron. When the firing rate of Neuron #2 is reduced, the likelihood of falling asleep increases. (b) When caffeine is present [dark gray (purple in the web version) diamonds], it inhibits the effects of adenosine. Because the firing rate of Neuron #2 is not inhibited, the likelihood for sleep is not increased with increased wakefulness.

loop. If individuals are sleep-deprived or groggy, then they may have a cup of coffee to increase their alertness. But as the day progresses, sleep debt builds up and the circadian system is not as active, such as in the early to mid-afternoon. Therefore, a common countermeasure is one of the caffeinated beverages mentioned previously. The effects of caffeine can take more than 8 h to wear off [9]. If a cup of coffee was consumed at around 3:00 pm, the effects may not subside until 11:00 pm that night. If the person attempts to go to sleep before that time, the consumed caffeine may delay the ability to fall asleep until the effects of caffeine completely wear off. Also note that everyone breaks down caffeine at a different rate. Further, each individual carries around their own level of sleep debt, which may be able to counteract the effects of caffeine. Therefore, individual results may vary on a day-to-day basis. With the long sleep latency from caffeine, the individual may not obtain the necessary sleep that night. This would lead to a sleep-deprived person that needs coffee to stay awake throughout the day. Therefore, judicious use of coffee can be very helpful to promote alertness and vigilance but with the knowledge that it cannot be used at all times.

Caffeine management can improve both sleep and wakefulness depending on how it is used. The primary recommendation, especially for someone who is having difficulties sleeping, is to not consume caffeine after lunch because it could interfere with one's sleep onset. It takes time for the caffeine to enter the system and inhibit the target receptor, usually 15–20 min for caffeine to begin to take effect. Therefore to increase alertness when driving, it is recommended to drink a cup of coffee and then take a brief nap (15–20 min) while the caffeine begins to have its effect on arousal [10]. For overnight shifts, it is recommended that one drinks caffeine throughout the shift so that the caffeine might have the alerting effect over the entire shift.

15.1.2 **Alcohol**

Alcohol is commonly used as a sleep aid, though it may not be as beneficial as one would hope. The idea of a "night cap" is a common occurrence. Alcohol can be both a stimulant at low doses and a sedative at higher doses. Typically when it comes to sleep, alcohol induces a two part response [11,12]. The first phase happens in approximately the first half of the night. During this period, individuals were able to fall asleep faster and they had increased amounts of slow-wave sleep with alcohol consumption prior to bedtime. Both of these changes are typically thought of as good, especially for those with insomnia. Though these initial effects may be beneficial, there is a trade-off with sleep later in the night. In the second half of the night, sleep becomes fragmented and disrupted as there is more wakefulness during that time period [13]. Thus, the resulting sleep fragmentation may result in inadequate sleep for that night. In addition, rapid eye movement (REM) sleep is suppressed throughout the entire night. REM sleep is more prominent in the second half of the night when sleep is more fragmented by alcohol consumption. REM suppression may be independent of the increased waking, as REM sleep is suppressed even in the first half of the night.

Alcoholism induces changes in sleep architecture. Sleep time is decreased and latency is increased on both nights that the person drinks as well as nights on which the person does not drink [11,12]. Even individuals who have been sober for up to 3 years exhibit less slow-wave sleep and increased REM sleep. Sleep architectural changes may predict whether an alcoholic will relapse [14,15]. Despite the inclination to use alcohol as a sleep aid, it imposes changes in the short-term as well as potential long-term changes in the stages of sleep.

15.1.3 **Nicotine**

Nicotine use disrupts sleep [16–18]. Whether it is the intermittent use or it is used on a consistent basis, the effects of nicotine are deleterious to sleep. Nicotine acts as a stimulant due to the activation of the nicotinic acetylcholine receptor in areas of the brain that activate wakefulness. Therefore, ingesting this compound will increase the activity of numerous different brain chemicals that are involved in initiating wakefulness. In two large population studies, active nicotine use was associated with sleep disruption, increased wakefulness, longer sleep latencies, increased NREM stage 1 sleep, and decreased amount of slow-wave sleep [19,20]. Interestingly, former smokers did not show differences in sleep parameters from those who had never smoked. One complication in interpreting these studies is that nicotine users often use more caffeine and alcohol than do control subjects [16]. Thus while one is an active smoker, one's sleep is disrupted, but once one has quit smoking and gone through a withdrawal period, sleep occurs similarly to the sleep of those who have never smoked.

"Social" nicotine users may also have their sleep disrupted. In laboratory experiments, nonsmokers who used a transdermal patch to deliver nicotine had decreased REM, increased stage 2 sleep, and earlier wake-up times compared to subjects with a placebo patch [20]. In another study, the authors documented a lower total sleep time, increased sleep latency, and decreased percentage of REM sleep over the night [20]. Thus, intermittent nicotine users also have disruptions in their sleep on the nights that they use nicotine. These people will often use nicotine close to bed time. In addition, this category of people will often combine nicotine with alcohol for a combined effect that can disrupt sleep throughout the night.

15.1.4 **Antihistamines**

Histamine plays many roles in the body. It is a molecule that helps two cells communicate to coordinate the body's actions. It is one of the molecules responsible for airborne allergic responses, which results in congestion, sneezing, and coughing. Antihistamines block the histamine H1 receptor that signals the induction of the allergic response. The first generation antihistamines, like doxylamine and diphenhydramine (active ingredient in Benedryl), block both the peripheral receptors, which are found outside of the central nervous system but also are able to penetrate the blood–brain barrier and block histamine H1 receptors in the brain [21]. There, these receptors help induce and maintain wakefulness. Therefore, drugs like diphenhydramine reduce the allergic reaction and wakefulness which help the person maintain sleep throughout the night [22]. Daytime administration may induce sleepiness and decrease cognitive performance [22–24]. For some, evening administration of this drug can lead to drowsiness or cognitive decrements the following morning. While the histamine H1 receptor is the primary target, diphenhydramine will also block cholinergic, serotonergic, and α-adrenergic receptors, all of which play a role in maintaining wakefulness [25]. In fact, diphenhydramine is the active ingredient in UniSom, the over-the-counter sleeping aid, as well as night-time versions of medications, such as ZzzQuil. Therefore, drug companies are increasingly using diphenhydramine as a sleep aid in numerous medications. Newer second generation antihistamines, such as loretidine (Claritin), fexofenadine (Allegra), and cetirizine (Zyrtec) do not cross the blood–brain barrier as easily and therefore do not have the same drowsiness effect on the person as the first generation drugs do. They may be a better choice under circumstances in which one does not want increased sleepiness or performance decrements [26].

15.1.5 Decongestant cough medications

Decongestants work by decreasing blood flow to the mucosal membranes and relieving the congestion. Common versions of this type of active ingredient include pseudoephedrine and phenylephrine. They can be found in many of the nondrowsy formulations of over-the-counter medications. These drugs typically activate α-adronergic receptors, reducing the dilation of blood vessels. Since they are often administered systemically, they activate the "fight-or-flight" system, otherwise known as the sympathetic system. The sympathetic system increases blood pressure, heart rate, pupil dilation, and dilates airways in an effort to protect us from danger. Activation of this system promotes wakefulness and decreases the likelihood of sleep. Among those active ingredients, pseudoephedrine shows the most likelihood to disrupt sleep [23,25,27]. Due to its role in synthesizing methamphetamines, it is harder to obtain medications containing pseudoephedrine and more medications consist of a substitute compound, phenylephrine. Phenylephrine (often abbreviated PE on over-the-counter pharmaceuticals), reduces congestion that accompanies a cold and other conditions, and also helps the person feel more alert through the day because of the stimulatory effects. Therefore, these pharmacological remedies can cause difficulties sleeping if taken close enough to bed, though phenylephrine appears less potent than pseudoephedrine [23]. Thus, when taking these medications, be aware of the time of day and how close you are to bed time so that sleep is not disrupted by relieving congestion. If necessary, one can take the night-time versions of these medications, which may help in maintaining sleep because they have the sedating antihistamine component.

15.2 PRESCRIPTION MEDICATIONS MEANT TO ALTER SLEEP AND SLEEPINESS

This portion of the chapter describes the effects and broad mechanisms of action for many of the common prescription medications that modulate sleep and wakefulness.

15.2.1 Benzodiazepines

Benzodiazepines were serendipitously discovered in the 1950s at Hoffman LaRoche, and have been among the most prescribed drug in Europe and the United States. This class of compound has been used to reduce anxiety, as an anticonvulsant, muscle relaxant as well as a sleeping aid, which is called a hypnotic. There are numerous variations of the original molecule that have slightly different effects and side effects, so no two are the same. The most recognizable might be Valium and Restoril, which are two of the most prescribed benzodiazepines. This class of drugs is associated with an increase in the inhibitory effects of γ-aminobutyric acid (GABA) by making the channel more open and active. Continued opening of this channel inhibits neuronal activity in the brain and reduces overall neuronal activity [28], which favors sleep. In fact, benzodiazepines typically reduce sleep latency and wake time while increasing total sleep time and improving subjective sleepiness [21,29]. Benzodiazepines can suppress slow-wave sleep and also may mildly suppress REM sleep [21]. Another effect involves changing people's perception about how well they have slept. In other words, there can be a misperception by people with sleep troubles that they are getting less sleep than objective measures demonstrate. Benzodiazepines appear to correct this misperception [21]. One of the downsides of benzodiazepines is that they have an increased level of dependence, and withdrawal from the drug can be difficult [30]. This can occur especially in people who have had a previous addiction. In practice, the nonbenzodiazepines are preferred for hypnotic purposes.

15.2.2 Nonbenzodiazipines somnogenics

Recently, a new class of hypnotic drugs has been developed that acts similar to benzodiazepines but with a shorter duration of action and with a decreased likelihood of misuse [28]. These drugs are collectively called the "Z-drugs" because of their names, zolpidem, zopiclone, zoleplon, and eszopiclone and are sold under some of the commonly prescribed hypnotics. They work similarly to benzodiazepines and increase the neuronal inhibition produced GABA receptors. But because they are a different class of compounds, they have a different profile and affect a smaller subset of GABA receptors. People who took the nonbenzodiazepines hypnotics had a better subjective and objective sleep rating compared to placebo, including sleep onset latency. Unlike benzodiazepines, zolpidem does not reduce slow-wave sleep [21]. Another benefit of the nonbenzodiazepines is that they do not exhibit the same likelihood of dependence as the benzodiazepines [31]. These compounds have become a very popular option for the short-term treatment of insomnia based on their pharmacologic and behavioral characteristics.

15.2.3 **Suvorexant**

This drug is one of the newest sleep aid treatments for insomnia. This is a sleep aid that reduces waking by blocking orexin receptors, which are responsible for inducing and maintaining wakefulness. Interestingly, the foundation for the development of this drug came directly from the research on the mechanisms that underlie narcolepsy. In these studies, the absence of the orexin protein resulted in the inability of the animal to maintain wakefulness throughout the wake period [32]. This was shown to be mediated by the orexin receptor [33]. These results formed the impetus to develop a drug that might promote sleep by reducing the wakefulness input. Clinical data show that this drug allows people to fall asleep faster, maintain sleep, and subjectively feel more refreshed after sleep [34]. Given that this drug is a novel approach, there will need to be continued research to understand what role this drug will play in the treatment regimen for insomnia. The development of Suvorexant represents the fulfillment of one of the promises of directed biomedical research. Here, a mechanistic study of narcolepsy and neuropeptides revealed a drug target that may benefit people with difficulties sleeping.

15.2.4 **Modanfinil/armodafinil**

Modafinil is used in cases of narcolepsy, people with shift work disorder, and those with excessive sleepiness from obstructive sleep apnea [35–37]. It is used to combat excessive sleepiness during the day without apparent negative effects on off-duty sleep [35]. This may be a better choice to combat excessive sleepiness than amphetamines because modafinil does not induce increases in blood pressure or heart rate. In addition, there is a lower likelihood of dependence, other hormones are not altered by modafinil, and modafinil may have a different and more tolerated method of alerting. Evidence suggests that modafinil modifies the dopamine reuptake system [38], though it has been shown to interact and alter multiple neurotransmitter systems [39]. The dopamine pathway is one of the major alerting pathways in the brain; thus at least one mode of action for modafinil is elevated levels of dopamine. There is some evidence that another arousal promoting neurotransmitter, norepinephrine is also increased, but this may be dose dependent. The benefits of this drug are that it reduces daytime sleepiness without many of the undesirable side effects of amphetamines.

15.2.5 Melatonin/melatonin receptor agonists

Melatonin is an endogenous hormone. With the onset of darkness in the evening, it is secreted from the pineal gland prior to one's typical bedtime and secretion stops in the morning around the time that one wakes up [40]. In sleep research, melatonin has been a readout of the endogenous circadian rhythms for an individual. One can now find melatonin at health food and nutrition stores and taken without a prescription. At low doses, melatonin has been used to shift the circadian clock to earlier times [41] and consequently it has often been observed that sleep onset latency is decreased [42]. In addition, there has been a documented increase in sleep duration, and melatonin can reduce the amount of waking over a night of sleep [40]. It is unclear if this effect is due to shifting the circadian clock or if the improved sleep is a direct effect of the compound. Melatonin works through two different melatonin receptors to exert its effects throughout the body. But it has recently been hypothesized that melatonin improves sleep independently from the effects on circadian rhythms. Melatonin has been tested in multiple patient populations. Because of the different experimental conditions, such as doses used and the time they were administered, it has been hard to draw firm conclusions on whether melatonin impacts sleep [43]. The most reliable effect was the reduction of sleep onset in patients. There was a documented increase in the ability to remain asleep and in sleep duration, but these were not as reliable as the effect of sleep onset. The key parameter for effectiveness is the timing of the dose. It appears that taking melatonin at a time when it is typically low will increase the likelihood that one will sleep [44]. Therefore, melatonin may be most useful for trying to induce sleep during the daytime and to reduce sleep onset latency for individuals that have a delayed bed time.

Based on the possibility that melatonin might aid with sleep problems, a drug that activates the melatonin receptors was developed to specifically activate these receptors, called Ramelteon. In trials, the largest effect was seen in the decrease in the latency to sleep. There were more modest increases in total sleep time and sleep efficiency among other variables [45,46]. Thus, it has been approved for the treatment of sleep onset latency. As with melatonin, it is unclear if Ramelteon directly modulates sleep or if the increase in sleep is carried out through the circadian system and permitting sleep earlier in the night. Ramelteon does seem to work in a similar way to melatonin and, along with melatonin, may provide relief for people with difficulties in sleep onset.

15.3 COMMON PRESCRIPTIONS THAT CAN ALTER SLEEP REGULATION

This portion of the chapter is to show that drugs prescribed for other reasons may also alter sleep.

15.3.1 β-Blockers

Drugs that address cardiovascular problems are commonly prescribed medications for the goals of reducing blood pressure and controlling cardiac problems. One such class of drugs is the β-blocker, which addresses cardiovascular issues by reducing heart rate and the force at which the heart pumps blood as well as by directly reducing the tension in the vessels. Generally, β-blockers reduce activity in the sympathetic nervous system, or the "fight-or-flight" response, but each individual compound has its own receptor selectivity with differing effects on sleep and wakefulness. As might be predicted from its mechanism of action, β-blockers are commonly associated with fatigue and sedation [23]. The β-blockers, carvedilol and labetalol, are associated with fatigue and somnolence, in part because they also block another set of closely related receptors, the α-adronergic receptors [47,48]. On the other hand, the brain chemistry of sleep is complex, and sometimes these compounds will inhibit other receptors in the system. For example, some β-blockers will inhibit a type of serotonin receptor, which will disrupt sleep [49]. Others, such as propranolol, metoprolol, and pindolol, are associated with increased wakefulness [23]. In addition, β-blockers may also inhibit the release of melatonin, which can alter sleep [50]. Each of these drugs is meant to treat cardiovascular problems, but each has a different effect on sleep and thus may be tolerated differently by each individual. Given the number of options, one should be able to find an option that is effective and well tolerated to minimize the impact on sleep.

15.3.2 Antidepressants

There is considerable overlap in the neurotransmitters that govern mental health and sleep. There are numerous drugs that attempt to treat depression, but many of these drugs directly modulate neurotransmitter systems that govern sleep and wakefulness, including cholinergic, histaminergic, serotinergic, and dopaminergic signaling. When active, these systems help keep you awake. The first generation of antidepressants, called the tricyclic antidepressants (TCAs), are known to act as hypnotics. Because of their inhibition of histaminic and cholinergic signaling as well as their block of serotonin reuptake, these can be very effective hypnotics [51]. They decrease sleep latency, increase total sleep time, and decrease REM sleep [43,52]. In fact, low-dose doxepin has been approved to treat insomnia. But these drugs are not all the same as other TCAs increase wakefulness because they have a different pharmacological profile. A closely related drug with a tetracyclic structure, trazodone, also has hypnotic effects. It is an inhibitor for a subclass of serotonin receptors, serotonin reuptake, and the histamine H1 receptor. This complex pharmacology has the effect of increasing sedation and improving subjective sleep quality in healthy individuals and patients with depression [51,53,54]. In younger subjects, the evidence is mixed on improved sleep metrics when assessed by brain activity. In older individuals, there is a documented improvement in sleep efficiency, total sleep time, and wakefulness with trazodone [52,55]. Trazodone is associated with an increase in slow-wave sleep and does not appear to suppress REM sleep [56,57]. For this reason, trazodone is used as a hypnotic as well as an antidepressant.

Another class of antidepressants, called the monamine oxidase inhibitors (MAOIs), can increase the levels of many of the neurotransmitters that increase wakefulness by inhibiting their breakdown by monoamine oxidase. Patients have complained of insomnia and that REM sleep is strongly suppressed in these patients and total sleep time is decreased [58]. Thus, these drugs may disrupt sleep in some patients. Another class of antidepressants are the selective serotonin reuptake inhibitors (SSRIs). They increase the levels of serotonin by blocking the enzyme that removes the serotonin. This inhibition keeps serotonin active for a longer period of time. This class of drugs can also block the uptake of dopamine and norepinephrine, as well. SSRIs may induce more sleep fragmentation and suppress REM sleep [51]. They will decrease total sleep time by increasing wake time. There is also more NREM Stage 1 sleep suggesting that sleep is lighter [51]. The antidepressant class of drugs shows the complex chemistry of sleep and how drugs that seem to have similar mechanisms may have differing effects on sleep in the patient. It also reveals how drugs taken for a very good purpose can have a problematic side effect and that there are other options if one particular drug does not work for the patient.

15.3.3 **Corticosteroids**

Corticosteroids are endogenous steroid hormones that help regulate physiological function, including the stress response, immune activity, ion balance in the blood, and metabolism. Cortisol, dexamethasone, and prednisone are all glucocorticoids used to treat conditions such as asthma, ulcerative colitis, autoimmune diseases, and to replace hormones in hormone-deficient individuals, among other ailments. Corticosteroids activate two different types of receptors, the glucocorticoid receptors, associated with the stress response, and the mineralocorticoid receptors, which govern ion levels in the body. Because activating the glucocorticoid receptor mimics the stress response, scientists have evaluated whether taking glucocorticoids alters sleep. Sleep is more likely to be affected when the drug is delivered throughout the body, such as, when it is ingested rather than when it is targeted to a specific tissue through an inhaler [23]. This type of administration level increases levels throughout the body and can mimic the stress response. Corticosteroids reduce sleep and increase wakefulness [59–61] and in particular suppress REM sleep [62]. Individuals taking prednisone report increased incidence of sleep disturbances and insomnia in a dose-dependent fashion [25]. Sleep disturbances are also seen in children taking prednisone. With dexamethasone, reductions in sleep and REM sleep and a delay in onset have been observed in healthy individuals [25]. The long-term use of inhaled corticosteroids does not appear to pose the same general risk, though reports of hyperactivity and insomnia have been noted [23].

The biochemistry of sleep is complex and involves numerous different chemicals and even more types of receptors and other proteins. All of these drugs attempt to help the individual and treat some problems that the person has. Ingesting these compounds may alter one's sleep either purposefully or inadvertently. This chapter is no substitute for consulting with a clinician to discuss the options, but it will help the individual to be informed about the effects these drugs can have on a person's sleep and potentially help design a treatment regimen that optimizes sleep opportunity, duration, and efficiency.

REFERENCES

[1] Staff MC. Caffeine content for coffee, tea, soda and more. 2014. Available from: http://www.mayoclinic.org/healthy-lifestyle/nutrition-and-healthy-eating/in-depth/caffeine/art-20049372

[2] Roehrs T, Roth T. Caffeine: sleep and daytime sleepiness. Sleep Med Rev 2008;12(2):153–62.

[3] Giles GE, et al. Differential cognitive effects of energy drink ingredients: caffeine, taurine, and glucose. Pharmacol Biochem Behav 2012;102(4):569–77.

[4] Administration, S.A.a.M.H.S. The DAWN Report: Update on Emergency Department Visits Involving Energy Drinks: A Continuing Public Health Concern. 2013; October 1, 2012. Available from: http://archive.samhsa.gov/data/2k13/DAWN126/sr126-energy-drinks-use.htm

[5] Benington JH, Kodali SK, Heller HC. Stimulation of A1 adenosine receptors mimics the electroencephalographic effects of sleep deprivation. Brain Res 1995;692(1–2):79–85.

[6] Satoh S, Matsumura H, Hayaishi O. Involvement of adenosine A2A receptor in sleep promotion. Eur J Pharmacol 1998;351(2):155–62.

[7] Porkka-Heiskanen T, et al. Adenosine: a mediator of the sleep-inducing effects of prolonged wakefulness. Science 1997;276(5316):1265–8.

[8] Huang ZL, et al. Adenosine A2A, but not A1, receptors mediate the arousal effect of caffeine. Nat Neurosci 2005;8(7):858–9.

[9] Penetar D, et al. Caffeine reversal of sleep deprivation effects on alertness and mood. Psychopharmacology (Berl) 1993;112(2–3):359–65.

[10] Reyner LA, Horne JA. Suppression of sleepiness in drivers: combination of caffeine with a short nap. Psychophysiology 1997;34(6):721–5.

[11] Ebrahim IO, et al. Alcohol and sleep I: effects on normal sleep. Alcohol Clin Exp Res 2013;37(4):539–49.

[12] Roehrs T, Roth T. Medication and substance abuse. In: Kryger M, Roth T, Dement WC, editors. Principles and Practice of Sleep Medicine. Saint Louis: Elsevier Saunders; 2010. p. 1512–23.

[13] Thakkar MM, Sharma R, Sahota P. Alcohol disrupts sleep homeostasis. Alcohol 2015;49(4):299–310.

[14] Gillin JC, et al. Increased pressure for rapid eye movement sleep at time of hospital admission predicts relapse in nondepressed patients with primary alcoholism at 3-month follow-up. Arch Gen Psychiatry 1994;51(3):189–97.

[15] Allen RP, et al. Slow wave sleep: a predictor of individual differences in response to drinking? Biol Psychiatry 1980;15(2):345–8.

[16] Lexcen FJ, Hicks RA. Does cigarette smoking increase sleep problems. Percept Mot Skills 1993;77(1):16–8.

[17] Soldatos CR, et al. Cigarette smoking associated with sleep difficulty. Science 1980;207(4430):551–3.

[18] Wetter DW, Young TB. The relation between cigarette smoking and sleep disturbance. Prev Med 1994;23(3):328–34.

[19] Zhang L, et al. Cigarette smoking and nocturnal sleep architecture. Am J Epidemiol 2006;164(6):529–37.

[20] Jaehne A, et al. Effects of nicotine on sleep during consumption, withdrawal and replacement therapy. Sleep Med Rev 2009;13(5):363–77.

[21] Mendelson W. Hypnotic medications. In: Kryger MH, Roth T, Dement WC, editors. Principles and Practices of Sleep Medicine. Saint Louis: Saunders Elsevier; 2010.

[22] Nolen TM. Sedative effects of antihistamines: safety, performance, learning, and quality of life. Clin Ther 1997;19(1):39–55. discussion 2-3.

[23] Schweitzer PK. Allergy/respiratory and cardiovascular drugs. Sleep Med Clin 2010;5:541–57.

[24] Shamsi Z, Hindmarch I. Sedation and antihistamines: a review of inter-drug differences using proportional impairment ratios. Hum Psychopharmacol 2000;15(S1): S3–S30.

[25] Schweitzer PK. Drugs that disturb sleep and wakefulness. In: Kryger MH, Roth T, Dement WC, editors. Principles and Practice of Sleep Medicine. Saint Louis: Elsevier; 2010. p. 544–60.

[26] Bender BG, et al. Sedation and performance impairment of diphenhydramine and second-generation antihistamines: a meta-analysis. J Allergy Clin Immunol 2003;111(4):770–6.

[27] Bye C, et al. A comparison of plasma levels of L(+) pseudoephedrine following different formulations, and their relation to cardiovascular and subjective effects in man. Eur J Clin Pharmacol 1975;8(1):47–53.

[28] Neubauer DN. New and emerging pharmacotherapeutic approaches for insomnia. Int Rev Psychiatry 2014;26(2):214–24.

[29] Nowell PD, et al. Benzodiazepines and zolpidem for chronic insomnia: a meta-analysis of treatment efficacy. JAMA 1997;278(24):2170–7.

[30] O'Brien CP. Benzodiazepine use, abuse, and dependence. J Clin Psychiatry 2005;66(Suppl. 2):28–33.

[31] Mendelson WB, et al. The treatment of chronic insomnia: drug indications, chronic use and abuse liability. Summary of a 2001 New Clinical Drug Evaluation Unit meeting symposium. Sleep Med Rev 2004;8(1):7–17.

[32] Chemelli RM, et al. Narcolepsy in orexin knockout mice: molecular genetics of sleep regulation. Cell 1999;98(4):437–51.

[33] Brisbare-Roch C, et al. Promotion of sleep by targeting the orexin system in rats, dogs and humans. Nat Med 2007;13(2):150–5.

[34] Michelson D, et al. Safety and efficacy of suvorexant during 1-year treatment of insomnia with subsequent abrupt treatment discontinuation: a phase 3 randomised, double-blind, placebo-controlled trial. Lancet Neurol 2014;13(5):461–71.

[35] Czeisler CA, et al. Modafinil for excessive sleepiness associated with shift-work sleep disorder. N Engl J Med 2005;353(5):476–86.

[36] Randomized trial of modafinil for the treatment of pathological somnolence in narcolepsy. US Modafinil in Narcolepsy Multicenter Study Group. Ann Neurol 1998;43(1):88–97.

[37] Randomized trial of modafinil as a treatment for the excessive daytime somnolence of narcolepsy: US Modafinil in Narcolepsy Multicenter Study Group. Neurology 2000; 54(5):1166–75.

[38] O'Malley MB, Gleeson SK, Weir ID. Wake-promoting medications. In: Kryger MH, Roth T, Dement WC, editors. Principles and Practice of Sleep Medicine. Saint Louis: Elsevier; 2010. p. 527–41.

[39] Minzenberg MJ, Carter CS. Modafinil: a review of neurochemical actions and effects on cognition. Neuropsychopharmacology 2008;33(7):1477–502.

[40] Zhdanova IV. Melatonin as a hypnotic: pro. Sleep Med Rev 2005;9(1):51–65.

[41] Burgess HJ, Revell VL, Eastman CI. A three pulse phase response curve to three milligrams of melatonin in humans. J Physiol 2008;586(2):639–47.

[42] Brzezinski A, et al. Effects of exogenous melatonin on sleep: a meta-analysis. Sleep Med Rev 2005;9(1):41–50.

[43] Krystal AD. Pharmacologic treatment: other medications. In: Kryger MH, Roth T, Dement WC, editors. Principles and Practice of Sleep Medicine. Saint Louis: Elsevier; 2010. p. 919–30.

[44] Scheer FA, Czeisler CA. Melatonin, sleep, and circadian rhythms. Sleep Med Rev 2005;9(1):5–9.

[45] Mayer G, et al. Efficacy and safety of 6-month nightly ramelteon administration in adults with chronic primary insomnia. Sleep 2009;32(3):351–60.

[46] Kuriyama A, Honda M, Hayashino Y. Ramelteon for the treatment of insomnia in adults: a systematic review and meta-analysis. Sleep Med 2014;15(4):385–92.

[47] McAinsh J, Cruickshank JM. Beta-blockers and central nervous system side effects. Pharmacol Ther 1990;46(2):163–97.

[48] Pearce CJ, Wallin JD. Labetalol and other agents that block both alpha- and beta-adrenergic receptors. Cleve Clin J Med 1994;61(1):59–69. quiz 80-2.

[49] Yamada Y, et al. Prediction of sleep disorders induced by beta-adrenergic receptor blocking agents based on receptor occupancy. J Pharmacokinet Biopharm 1995;23(2):131–45.

[50] Stoschitzky K, et al. Influence of beta-blockers on melatonin release. Eur J Clin Pharmacol 1999;55(2):111–5.

[51] Krystal AD. Antidepressant and antipsychotic drugs. Sleep Med Clin 2010;5:571–89.

[52] Buysse D. Clinical pharmacology of other drugs used as hypnotics. In: Kryger MH, Roth T, Dement WC, editors. Principles and Practice of Sleep Medicine. Saint Louis: Elsevier; 2010. p. 492–509.

[53] Montgomery I, et al. Trazodone enhances sleep in subjective quality but not in objective duration. Br J Clin Pharmacol 1983;16(2):139–44.

[54] Saletu-Zyhlarz GM, et al. Insomnia in depression: differences in objective and subjective sleep and awakening quality to normal controls and acute effects of trazodone. Prog Neuropsychopharmacol Biol Psychiatry 2002;26(2):249–60.

[55] Le Bon O, et al. Double-blind, placebo-controlled study of the efficacy of trazodone in alcohol post-withdrawal syndrome: polysomnographic and clinical evaluations. J Clin Psychopharmacol 2003;23(4):377–83.

[56] Maixner S, et al. Effects of antipsychotic treatment on polysomnographic measures in schizophrenia: a replication and extension. Am J Psychiatry 1998;155(11):1600–2.

[57] Ware JC, Pittard JT. Increased deep sleep after trazodone use: a double-blind placebo-controlled study in healthy young adults. J Clin Psychiatry 1990;51 (Suppl.):18–22.

[58] Mayers AG, Baldwin DS. Antidepressants and their effect on sleep. Hum Psychopharmacol 2005;20(8):533–59.

[59] Bailey WC, et al. Characteristics and correlates of asthma in a university clinic population. Chest 1990;98(4):821–8.

[60] Chrousos GA, et al. Side effects of glucocorticoid treatment. Experience of the Optic Neuritis Treatment Trial. JAMA 1993;269(16):2110–2.

[61] Lozada F, Silverman S Jr, Migliorati C. Adverse side effects associated with prednisone in the treatment of patients with oral inflammatory ulcerative diseases. J Am Dent Assoc 1984;109(2):269–70.

[62] Born J, et al. Differential effects of hydrocortisone, fluocortolone, and aldosterone on nocturnal sleep in humans. Acta Endocrinol (Copenh) 1987;116(1):129–37.

16

Creating a fatigue risk management system (FRMS)

16.1 CALL FOR FATIGUE RISK MANAGEMENT SYSTEMS (FRMS)

Excess workplace fatigue can be a major risk to safe production operations and increased work hours contribute to these risks. In the past, some thought that simply placing limits on work hours could adequately address the issue. However, in recent years a consensus has emerged that a better approach to mitigate fatigue related risks is through a comprehensive fatigue risk management system (FRMS). As with any other safety management systems, a FRMS needs to be integrated with other policies and procedures. Everyone involved has a role in recognizing the importance of workplace fatigue risk mitigation and actively working to support the goals of the FRMS. Transportation industries were leaders in advocating and developing programs related to fatigue management and countermeasures. Recently, these efforts have extended into other industries including chemical processing [1].

After the Texas City accident (see chapter: The Consequences of Fatigue in the Process Industries), the US Chemical Safety and Hazard Investigation Board (CSB) recommended the development of fatigue prevention guidelines that addressed limits on hours of work, shift schedules, and similar changes with the goal of managing fatigue in the workplace. ANSI/API Recommended Practice 755 – Fatigue Prevention Guidelines for the Refining and Petrochemical Industries [2] was industry's response to that recommendation. It relies on the use of FRMS to address the hazard of human fatigue in the processing industry. The guideline calls for the development of procedures, information, and training for employees and employers, and monitoring to understand and manage fatigue in the workplace.

Another FRMS guidance document is *Performance Indicators for Fatigue Risk Management Systems* [3]. It was written by the Fatigue Risk Management Task Force of the International Association of Oil & Gas Producers

(IOGP, formally known as OGP) and the global oil and gas industry association for environmental and social issues (IPIECA). In 2012, the report issued its OGP Report Number 488. This document proposes performance indicators that can be used in developing and monitoring an FRMS. In it the authors explain than while limiting workers' hours has been the traditional approach to managing the risks associated with human fatigue that is not sufficient to manage the problem. They note that FRMS provide greater operational flexibility while being more comprehensive and scientifically-based.

The US Department of Transportation (Pipeline and Hazardous Materials Safety Administration also called for FRMS in *49CFR: 192.631* and *195.446 Control room management*). These regulations apply to control room operators of pipeline facilities. They call for a written plan that includes a fatigue mitigation plan. Specifically, it states an FRMS should include: (1) work schedules that allow for sufficient off-duty time to achieve 8 h of continuous sleep, (2) education on fatigue mitigation strategies, (3) training to recognize the effects of fatigue, and (4) maximum limits of pipeline controllers' hours of service (HOS) [4,5].

The Federal Aviation Administration (FAA) has long had regulations concerning duty-hour and crew rest hours for pilots. In recent years, the FAA has also required dispatchers, schedulers, and those working in operational control be covered in a FRMS. The FAA requirements can be found in *14CFR 117.7, 121.473*, and *121.527 Fatigue risk management system*. The FAA requires that a FRMS include management policies, training/education, a fatigue reporting system, an incident reporting process, and a performance evaluation plan. As with other government regulatory agencies the FAA's goal is to improve the level of safety against the risk of fatigue-related accidents.

The Federal Railroad Administration (FRA) via the *National Rail Safety Action Plan Final Report 2005–2008* highlighted the problem of fatigue among railroad operating employees [6]. The FRA called for improved crew scheduling to manage the risk of fatigue in the railroad industry. The FRA [7] acknowledges that an approach that limits HOS is necessary but not sufficient to prevent operator fatigue and called for proactive fatigue risk management programs that balance the amount of work, when it is performed, and other factors.

In their *Guidance Statement on Fatigue Risk Management in the Workplace* [8], the American College of Occupational and Environmental Medicine (ACOEM) addressed the needs of safety-sensitive operations such as transportation, health care, and energy industries to manage fatigue in the workplace. Their recommendations are not industry specific, but their focus is on organizations with workers working around the clock on extended shifts. As medical professionals they bring an important perspective to the call for FRMS in industries beyond transportation.

16.2 PURPOSE OF AN FRMS

Alert, well-rested employees are key to safe operations in the processing industry. Organizations with policies that encourage employees' well-being and supportive work environments can promote alertness and reduce accidents. The processing industry has begun to realize that managing HOS is not enough. Managing the risks associated with fatigued workers requires a multifaceted approach. An effective FRMS should include sections that address:

- Roles and responsibilities of the worker and management;
- Training on sleep, sleep disorders, and fatigue counter measures;
- Risk assessment and risk mitigation;
- Work schedules;
- Work environment;
- HOS guidelines;
- Accident/incident investigation; and
- Monitoring and assessing the FRMS.

An FRMS should be developed with the goal of effectively managing fatigue-related risks. The organization must make a commitment to adhere to the fatigue management plan even if it impacts production in the short-term. As with any safety system the goal is risk identification, assessment, mitigation, and monitoring. Once the plan is developed it must be implemented, monitored, assessed, and modified as shown in Fig. 16.1.

16.3 ROLES AND RESPONSIBILITIES

It is common for safety regulations to list roles and responsibilities for both the worker and the organization. Managing fatigue risk is unique in that a significant portion of the employees' responsibility is outside of the workplace beyond the control of management. Employees have a duty to manage their sleep and their overall well-being. The only remedy for sleep deprivation is to sleep. Sleep is a complex process; it varies person-to-person, and is influenced by many things [9]. Employees should accept the responsibility to manage their sleep and their personal sleep schedule. They should accept the responsibility to report when they are too fatigued or otherwise unfit to safely perform the work. If they observe coworkers who are overly sleepy or

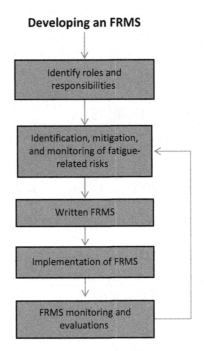

Developing an FRMS

Identify roles and responsibilities

Identification, mitigation, and monitoring of fatigue-related risks

Written FRMS

Implementation of FRMS

FRMS monitoring and evaluations

■ **FIGURE 16.1 Steps in the development of an FRMS.**

performing poorly regardless of the cause they should speak up. Employees should seek medical help when they are suffering from sleep disorders or otherwise unable to sleep sufficiently.

In managing the risks associated with human fatigue the workers have an unusual role. They are typically the most aware of the problem and the best able to fix. Yet, they are often faced with pressure both internally and externally to "tough it out." Ensuring they have sufficient sleep can mean missing out on extra work hours and the financial incentives associated with overtime and additional shifts. There can be peer pressure to work long hours. The organization should schedule work to provide sufficient time for rest, provide fatigue risk management training, and manage the FRMS, but getting the necessary sleep is up to the employee.

Table 16.1 shows the roles of the organization, supervision, and the employees. An FRMS will work only if all three groups are committed to the program and take responsibility for their portion. Upper management of the organization must commit to the importance of fatigue risk management. This can be done in a letter similar to the example in Fig. 16.2. A focus on fatigue risk management can be a paradigm shift in industries that are accustom to doing more with less and focusing on the bottom line.

Table 16.1 Responsibilities in an FRMS

Organizational Responsibilities	Supervisor Responsibilities	Individual Responsibilities
Develop an FRMS	Determine jobs/workers that should be covered by FRMS	Comply with FRMS requirements
Provide policies and training related to work schedules and fatigue-related risks	Ensure employees receive proper FRMS training	Take a proactive role in learning about fatigue-related risks and managing their own well-being Obtain adequate sleep, maintain overall health, and be fit to work
Assess compliance with the FRMS	Insure work schedule provide adequate opportunity for rest	Report any fatigue-related symptoms
Investigate fatigue-related accidents as needed	Report any fatigue-related risks and/or incidents	Report any fatigue-related safety risks

FATIGUE RISK MANAGEMENT SYSTEM (FRMS)
IMPLEMENTATION GUIDE FOR OPERATORS

EXECUTIVE LETTER

Dear Colleagues,

Air travel continues to be the safest means of transportation, but that does not allow for us to become complacent. We continually strive for improvements in our industry safety record, which is a testament to our ongoing commitment to safety.

Fatigue Risk Management Systems (FRMS) continues the move from prescriptive to performance based regulatory oversight. As in Safety Management Systems (SMS), FRMS strives to find the realistic balance between safety, productivity and costs in an organization, through collection of data and a formal assessment of risk.

Traditionally, crewmember fatigue has been managed through prescribed limits on maximum flight and duty hours, based on a historical understanding of fatigue through simple work and rest period relationships. New knowledge related to the effects of sleep and circadian rhythms provides an additional dimension to the management of fatigue risks. An FRMS provides a means of adding this safety dimension, allowing operators to work both safer and more efficiently.

This FRMS Implementation Guide for Operators is a significant milestone. It marks the successful collaboration between IATA, IFALPA and ICAO to jointly lead and serve industry in the ongoing development of fatigue management, using the most current science. The input of these three organizations has ensured that this document presents a scientifically-based approach that is widely acceptable to the operators and the crew members who will be using it. It also offers this We are extremely proud to mutually introduce this FRMS Implementation Guide for Operators, which will contribute to the improved management of fatigue risk, and to ultimately achieve our common goal of improving aviation safety worldwide.

Guenther Matschnigg
Senior Vice President
Safety, Operations &
Infrastructure, IATA

Nancy Graham
Director
Air Navigation Bureau
ICAO

Don Wykoff
President
International Federation of
Air Line Pilots' Associations

■**FIGURE 16.2 Example management support letter.** *(Source: From: http://www.iata.org/ publications/Documents/FRMS%20Implementation%20Guide%20for%20Operators%201st%20 Edition-%20English.pdf).*

The organization's management is responsible for developing the FRMS. This should be done in accordance with industry guidelines and government regulations. They must provide adequate resources to create a FRMS specific to their organization. Policies and procedures concerning fatigue should be developed in consultation with workers. They have practical understanding of work processes and will be the ones implementing the FRMS. Once the FRMS is approved, training should be developed for (1) managers who will be responsible for following the procedures including scheduling work shifts and proper training and for (2) the workers covered by the FRMS. Considering the potential of fatigue as a cause in accident investigations should added to existing procedures in the safety department (this is discussed in more detail in chapter: Accident Investigation). As with any safety policy, management should ensure the policy is being followed at the operational level, that the associated records are being kept, and that the policy is evaluated and updated as needed.

Supervisors need to understand the dangers of overly tired workers and respect the potential for critical errors and mistakes when workers are not well-rested. They have distinct responsibilities to ensure a FRMS is successfully implemented. Fundamentally, they should schedule work to provide sufficient time for rest, provide fatigue risk management training, and manage the FRMS. This requires a familiarization with the organization's FRMS. Determining which jobs should be covered by the policy. This is often based on whether the work is performed on multiple shifts. Supervisors should be held accountable for having their workers attend FRMS training to introduce the policy and sleep related issues. The supervisors should take more extensive training to also understanding work schedules and polices related to HOS. Supervisor training should also address reporting requirements and resources available within the organization.

16.4 FRMS IMPLEMENTATION

The FRMS should be integrated into the organization's existing operational plans, safety policies, and training programs. The FRMS should be:

- Specific to the site;
- Developed through consultation with stakeholders;
- Available to employees;
- Discussed regularly; and
- Reviewed periodically.

The fatigue management plan should be fully documented. The documentation should include:

- Management's commitment to managing fatigue;
- Roles and responsibilities within the FRMS;
- Risk that have been identified;
- Risk mitigation and implementation strategies;
- Required training on sleep, fatigue-related risks, and recognizing the effects of fatigue;
- HOS policies;
- Schedule rotation policies;
- Record keeping requirements
- Support systems (hours-of-work monitoring, employee assistance programs, medical resources, etc.);
- Process to monitor and revise the plan.

Key questions to be addressed in any FRMS

What work hours must be included in the calculation of duty hours for fatigue mitigation consideration? The answer may be more extensive that some would expect. HOS should include not only the time when an individual is performing his or her operational tasks. The time spent on shift change should be included, such as updating shift logs, hand-off conversations, and the time required to store equipment and leave the facility. In some situations the time required to travel to and from the worksite maybe consider in the hours of work or at least time that is not available for the individual to sleep. Work that is required but does not occur regularly should not be overlooked in the monitoring of work hours. This could include on-call duties, call-outs, incident response, training, and other work-related events.

Whose hours should be tracked in the FRMS? Shift workers, production operators, and other workers with well-defined duties are easy to identify for inclusion in a FRMS tracking system. However, there are often other workers with less well-defined duties that can be overlooked that should be considered for inclusion in the FRMS. This can include supervisors and individuals who may be called on to cover short-term operational needs (individuals who have maintained the work certification but it is not their primary duty). Maintenance personnel should be included in the FRMS if they are called into work outside their normally scheduled work hours. Other employees who support production operations could warrant inclusion if their work hours are unpredictable and included extensive overtime.

What minimum time should be scheduled between shifts to provide sufficient time to achieve 8 h of continuous sleep? The opportunity for 8 h of continuous sleep between shifts is recommended. To achieve this, a worker often needs at least 10 continuous hours of off-duty time to allow for commutes and other personal activities prior to going to sleep or after waking up. The CSB investigation into the sleep hour of operators involved in the Texas City accident, confirmed the tendency of workers to sacrifice sleep for time with family and attending to personal needs.

How can an organization dictate what employees do in their time away from work? An FRMS attempts to reduce the risks associated with fatigued workers and setting the expectation that workers are fit for work during their work shifts. An FRMS is concerned with educating people about sleep and fatigue-related risks and providing workers with enough time away from work to obtain adequate sleep. Employees are expected to manage their behavior to be ready to work. This is no different to showing up sober or taking time off when you are sick.

What do you say to a worker who claims he or she only needs a few hours of sleep per night? Sleep researchers have found that most people need between 7–9 h of sleep per night. Only a small percentage of people can be healthy and perform well with significantly less sleep night after night. People may have gotten used to functioning on less sleep than they need, but that does not mean that they are functioning as well as they should. An example is poor sleep due to a medical condition such as sleep apnea. A worker might want to use a sleep diary and monitor how they feel when they have varying amounts of sleep. A couple of questions may help distinguish between bravado and biology. Ask if the person can make it through the day without a stimulant. Another question would be to ask if they sleep in on their days off. If they need caffeine or extended sleep, then they likely are not so-called "short sleepers." Another possibility would be to have a sleep study performed to determine the amount and quality of sleep the person is getting.

16.5 TRAINING

Safety training gives people the knowledge to understand and manage risk. Employees in the hazardous industries experience training for a variety of hazards. Safety professionals know that training is not the only component of an effective safety strategy but is an essential one. Training gives the workforce the skills they need to work in hazardous situations. Fatigue management can present some unusual challenges to the trainer. Everyone has slept, yet they may not have much understanding of the biology of sleep. An employee can be suffering from a sleep disorder or inadequate sleep and not know it. There can be a sense of pride associated with working long hours and getting little sleep.

The training must not only provide appropriate information about the fatigue hazards and risks in the workplace but also convince the employees of the importance of sleep. Workers should be informed about the fatigue management plan and how it will be implemented. The training should include:

- Basic sleep, circadian rhythms, and fatigue;
- Warning signals of fatigue and potential effects of fatigue;
- Strategies for quality, restorative sleep;
- Factors that influence fatigue (ie, sleep disorders, medicines);
- An overview of the organization's FRMS; and
- An explanation of how the FRMS will be implemented.

Additional training should be provided to supervisors that includes:

- Influence of staffing levels and schedules of employee fatigue;
- Warning signs of excessive fatigue; and
- Policies concerning hours of service and work schedules.

16.6 HOURS OF SERVICE LIMITS

An FRMS needs to specify HOS limits. The goal of HOS limits is to prevent workers working at levels that will result in chronic sleep debt. The FRMS should set the base scheduled hours. This could be as simple as 40 h per week or maybe different work periods. The FRMS address limits based on shift lengths (eg, 8, 10, or 12 h) and exceptions beyond those. The HOS limits can be based on industry guidelines or government regulations. An example of HOS limits is present in Table 16.2.

Table 16.2 Sample Hours of Service Limits

Factor	8 h Shift	10 h Shift	12 h Shift
Maximum work set (consecutive shift without a day off)	10 shifts—normal operations 19 shifts—outages	9 shifts—normal operations 14 shifts—outages	7 shifts—normal operations 14 shifts—outages
Minimum hours off after a work set	36 h for day shifts 48 h for night shifts	36 h for day shifts 48 h for night shifts	36 h for day shifts 48 h for night shifts
Maximum holdover period	4 h	2 h	2 h
Maximum extended shift (to be used rarely, only when necessary)	16 h	16 h	18 h
Minimum hours off after an extended shift	8 h	8 h	10 h
Maximum extended hour shifts per work set	2 h per work set	2 h per work set	1 h per work set

16.7 FRMS RESOURCES

Most organizations will have resources, both internally and externally, that can be beneficial to those developing or implementing a FRMS. A section of the FRMS can be customized to the company or location. An office of occupational health and safety within an organization might have expertise related to the workplace environment. Often they have meters that can measure sound, lighting, or air quality. If there are problems in these areas, they can often help in the process of improving the environmental conditions in the workplace. An industrial engineering department can assist in assessing work tasks, workload, or ergonomics. Large companies often provide free or inexpensive employee assistance programs. Workers can confidentially speak to a trained professional about stress or other issues that maybe affective their overall health or the quality of their sleep. Employee wellness programs are becoming more common. Stress management, weight loss programs, smoking cessation, substance abuse programs, and exercise programs are examples of resources that could benefit a worker seeking to improve their well-being including getting a better night's sleep.

Externally, other resources may be worth spotlighting in the FRMS. A local sleep clinic that performs diagnostics testing for sleep disorders would be a logical resource to share with both managers and employees covered by the FRMS. A wide variety of products are available that have been designed to help people get a better night's sleep. Shift workers who are trying to sleep during daylight hours might be interested in blackout curtains or eye masks. Other products include white noise sound machines, relaxation CDs, and specially designed pillows. There is always the concern that some of these products may be more marketing than scientifically sound devices, but someone struggling with poor sleep and excessive fatigue might want to try them.

16.8 ASSESSING AN FRMS

Creating a FRMS is only one step in the process of managing the risk of human fatigue. The system should be assessed periodically and changes made as needed. The following are questions adapted from the IPIECA's *Performance Indicators for Fatigue Risk Management Systems* [10].

- Has a single point of accountability be identified for the FRMS?
- Does the FRMS comply with other organizational polices and/or collective bargaining agreements?
- Have the jobs and workers covered by the FRMS been specifically identified?

- Has the potential impact of fatigued workers been considered?
- Does the work schedule allow for reasonable opportunity for each work to have 7–8 h of sleep every 24-h period?
- Are the work schedules reviewed systematically to allow for effective fatigue management?
- Are the differences between planned work hours and actual work hours reviewed?
- How often do actual work hours exceed recommended work hours? Are corrective actions taken to resolve this situation?
- What is the policy to manage additional work hours (ie, call-outs, startups)? Is being followed?
- Has the work environment (lighting, temperature, work station design, etc.) been evaluated with respect to work fatigue?
- Has the work task (type of work, variety of work, breaks, etc.) been evaluated with respect to work fatigue?
- Are workers informed about sleep, their personal responsibilities, sleep disorders, and the FRMS?
- Have the training requirements with the FRMS been met?
- How are potential fitness-for-duty issues managed for safety-sensitive duties?
- Is fatigue considered as a potential factor in accident investigations?
- Has the program monitoring process for the FRMS been followed?

Records should be kept when the FRMS is updated and when it is due for another review. Any FRMS is only as good as its implementation. After it is developed, management must insure it is followed. For some organizations a new FRMS will represent a major change in operations and can require a significant cultural change within the organization.

16.9 FRMS QUALITY ASSURANCE QUESTIONS

The New South Wales Mining and Extractive Industry Health Management Advisory Committee has written a detailed Fatigue Management Evaluation Manual [11] that details how an FRMS can be evaluated. One of the tools included in the document is a list of quality assurance questions that can be asked of various stakeholders during a FRMS audit. The following are some of the questions they suggest:

For managers and supervisors:

- What do you know of fatigue and how it is managed here on site?
- Are you aware if your site has a FRMS?
- What does the FRMS cover?

- What responsibilities do you have under the FRMS?
- What training have you received related to the FRMS?
- What are the HOS limitations for your job?
- How is the success of the FRMS monitored?
- Have you or someone else self-reported as being fatigued?

Additional questions for direct supervisors:

- How is fatigue assessed in individuals at this site?
- Have you ever assessed anyone as being fatigued?
- How are you assured of management support when you make a decision regarding fatigue management reporting or assessment?
- How do you ensure that the FRMS is promoted and understood by those it covers?
- From your knowledge, how often has fatigued been identified as a potential contributing factor in an incident report?
- What monitoring of the FRMS are you aware of?

Questions for workers:

- What do you know of how fatigue is managed here on site?
- What responsibilities do you have as a worker under the FRMS?
- What training have you received related to the FRMS? When did you receive it?
- What are the HOS limitations for your particular job?
- How is fatigue assessed in individuals at this site?
- Are you aware of anyone being assessed as fatigued or who needs to be?
- Have you ever self-reported as being fatigued? What happened?
- What is the biggest issue surrounding fatigue at this site? Is it being managed well?

These questions are a starting point in the evaluation process. It is important the FRMS is monitored and assessed after it is implemented. It is key to determine how, or even if, the FRMS is being used. Is it fundamentally incorporated in the daily operations at the facility or it is a policy graveyard and never used and having no influence on operations? Does the FRMS reduce the risks associated with fatigued workers? Does the FRMS decrease the probability that sleep-deprived workers are working in the facility?

REFERENCES

[1] Moore-Ede M. Evolution of Fatigue Risk Management Systems: The "Tipping Point" of employee fatigue mitigation. CIRCADIAN® White Paper. www.circadian.com/pages/157_white_papers.cfm; 2009.

[2] ANSI/API. Fatigue risk management systems for personnel in the refining and petrochemical industries. American National Standards Institute (ANSI)/American Petroleum Institute (API) Recommended Practice 755. 1st ed., April 2010.

[3] http://www.iogp.org/pubs/488.pdf

[4] http://www.gpo.gov/fdsys/granule/CFR-2011-title49-vol3/CFR-2011

[5] http://primis.phmsa.dot.gov/crm/docs/SGA_July2011.pdf

[6] Federal Railroad Administration, National Rail Safety Action Plan Final Report 2005-2008 http://www.motleyrice.com/sites/default/files/documents/PI_WD/nps49-102210-11.pdf

[7] Federal Railroad Administration: Office of Research and Development 2006 The Railroad Fatigue Risk Management Program at the Federal Railroad Administration: Past, Present and Future.

[8] ACOEM 2012. Fatigue risk management in the workplace. J Occup Environ Med, 54, p. 231–258. American College of Occupational and Environment Medicine (ACOEM) Guidance Statement. ACOEM Presidential Task Force on Fatigue Risk Management: Lerman SE, Eskin E, Flower DJ, George EC, Gerson B, Hartenbaum,N, Hursh SR Moore-Ede M.

[9] http://onlinepubs.trb.org/onlinepubs/tcrp/tcrp_rpt_81.pdf Toolbox from Transit Operator Fatigue

[10] IPIECA's Performance Indicators for Fatigue Risk Management Systems 2012.

[11] NSW Mining and Extractive Industry Health Management Advisory Committee (HMAC), *Fatigue Management Evaluation Manual*. http://www.resourcesandenergy.nsw.gov.au/__data/assets/pdf_file/0003/472260/Fatigue-eval-man-310713.pdf; 2013.

Accident investigation

17.1 INVESTIGATING ACCIDENTS, INCIDENTS, AND NEAR MISSES

Safety professionals and regulatory bodies encourage the investigation of accidents, incidents, and near misses. Investigations after an accident with injuries, loss of life, or major financial damage are commonly conducted. These can be done by outside agencies including the Occupational Safety and Health Administration (OSHA), law enforcement, and insurance companies, as well as internal investigations. Similarly, investigations should also be performed for incidents that had less severe consequences, and near misses which might not have had consequences. These investigations can identify issues that, if left uncorrected, could result in severe accidents in the future. The same is true of near misses or close calls that were narrowly averted but had the potential of significant damage. A greater number of near misses or undesirable events may occur than accidents, but in some organizational cultures they are not investigated as often. Underreporting and lack of investigation can be caused by:

- Fear of disciplinary actions;
- Fear of embarrassment;
- Lack of management commitment to safety;
- Lack of understanding about when to investigate;
- Difficult process to report and/or investigate incidents; and
- Negative consequences tied to incident rates.

An organization that does not effectively investigate accidents and incidents has major safety issues beyond just managing fatigue risk. This should be addressed before attempting to incorporate components of a fatigue risk management system (FRMS) into the organization's accident investigation process. OSHA has regulations guiding accident investigations, these can be found online at the OSHA.gov website. Regulation *29CFR 1960.29* is a good starting point for developing an accident investigation program. The agency also provides training and sample forms. Many other countries' governments and regulatory agencies have similar requirements and resources;

Table 17.1 Accident Investigation Process

Phase	Activities	Results
Accident investigation preparation	■ Develop policies, procedures, and documents for accident investigation ■ Gather resources for investigations (ie, forms, measurement equipment, cameras) ■ Select and train individuals to conduct investigations	■ An organization that is prepared to quickly and accurately investigate accidents and incidents at its facilities
Accident response	■ Ensure first aid, medical treatment, and emergency response are provided as needed ■ Secure the scene ■ Gather information ■ Interview witnesses ■ Document the incident with photos, reports, and forms	■ A report that details what happened, who was involved, potential causes, and the consequences of the accident or incident
Analysis	■ Safety professionals and subject matter experts review the reports ■ Answer the question "why" by identifying cause(s) and contributing factor(s) ■ Determine corrective actions	■ A report that contains recommendations and corrective actions that should be taken to avoid a reoccurrence of the event
Corrective action	■ Implement recommendations and corrective actions ■ Track completion ■ Evaluation effectiveness of changes ■ Disseminate information with others	■ Modified policies and procedures ■ Safer processes

for example, the United Kingdom has the Health and Safety Executive [1]. The Center for Chemical Process Safety of the American Institute of Chemical Engineers has published a book, *Guidelines for Investigating Chemical Process Incidents*, that provides a detail instruction on the topic [2].

The process of accident investigation is detailed in Table 17.1. As shown in the table, preliminary work should be done before an accident occurs. It is important for an organization to prepare. Once an accident, incident, or near miss occurs it is important that it is reported and investigated promptly. This is followed by writing and submitting reports concerning the event. The most important portion of the investigation is unfortunately also the most often overlooked part; insuring measures are taken to prevent a reoccurrence not only at the specific facility where the accident occurred but organization-wide. In some situations, findings might justify dissemination beyond the organization to the industry and other interested parties.

17.2 CONSIDERING HUMAN FACTORS IN AN INVESTIGATION

Countless accident investigations have concluded that the accident occurred because of *operator error*. This may be an easy conclusion to reach. It may even be a satisfying conclusion to reach because the solution is to fire the individual involved or add more training. However, operator error is often a symptom of larger underlying problems. The Health and Safety Executive, a British government agency, lists factors that can be underlying these human error findings in their *Inspector Toolkit* [1].

Job factors:

- Illogical design of equipment and instruments;
- Constant disturbances and interruptions;
- Missing or unclear instructions;
- Poorly maintained equipment;
- High workload; and
- Noisy and unpleasant working conditions.

Individual factors:

- Low skill and competence levels;
- Tired staff;
- Bored or disheartened staff; and
- Individual medical problems.

Organization and management factors:

- Poor work planning, leading to high work pressure;
- Lack of safety systems and barriers;
- Inadequate responses to previous incidents;
- Management based on one-way communications;
- Deficient coordination and responsibilities;
- Poor management of health and safety; and
- Poor health and safety culture.

An effective accident investigation team that is not satisfied with *operator error* can use these potential factors to drill deeper during the investigation and identify larger problems that can be corrected resulting improved safety.

17.3 FATIGUE AS A CONTRIBUTING FACTOR IN ACCIDENTS

An organization that is striving to manage the risk of human fatigue should evaluate whether fatigue or a failure with the FRMS could have contributed to the accidents, incidents, and hazards that they investigate. This provides an opportunity to learn how to improve processes, procedures, training, and culture to reduce the hazards associated with fatigued workers, reducing the risks in the future. To accomplish this, all investigations must explicitly include methods to assess this concern. At the minimum, forms reporting an accident should include the time of day when the accident occurred and the point during the employee(s) work shift (ie, prior to work shift, during normal work shift, overtime, callout, etc.). This basic information can provide some initial indicators that worker fatigue should be considered further in the investigation.

Those conducting accident and incident investigation should be trained on human fatigue, the effects of sleep loss, and the organization's FRMS. Their investigations into incidents in which worker fatigue is a potential contributing factor should consider both the workers and the work:

■ Continuous hours of wakefulness;
■ A 72 h timeline prior to the accident;
■ Short-term and long-term sleep loss among those involved;
■ Sleep disorders among those involved;
■ FRMS training;
■ Whether work schedules comply with FRMS requirements;
■ Work environment;
■ Presence of any alertness strategies; and
■ Work task requirements.

The investigator should attempt to determine if any fatigue-related factors were present at the time of the incident. The severity of these fatigue factors and the potential of these contributing to the incident should also be explored. The investigation should also consider the organization's FRMS. Is the policy sufficient? Was it being followed at the time of the accident or incident? Are changes to the FRMS needed? Is FRMS training being done? Is it effective? The work task and work environment should be examined to see if it contributes to workers feeling sleepy. Can changes be made to increase alertness levels, such as periodically walking to other areas or taking breaks?

The Chemical Safety Board (CSB) conducts detailed investigations of accidents in the chemical industry. Their investigation of the BP Texas City explosion found numerous problems that contributed to the accident. Acute sleep loss and cumulative sleep debt were found to have likely contributed to the accident [3]. The full report is available on the agency's website at CSB.gov. While fatigue was not the primary focus of the investigation, it was one of the first chemical accident investigations that documented lack of sleep as a contributing factor. There are accident investigation from other industries that have focused primarily on fatigue and lack of sleep that are excellent examples of evaluating fatigue as an accident cause.

17.4 SAMPLE NTSB FATIGUE-RELATED ACCIDENT INVESTIGATION

On Jun. 26, 2009 a truck–tractor semitrailer was involved in a rear-end collision on Interstate 44 near Miami, Oklahoma. The National Transportation Safety Board (NTSB) investigated the accident and released a report [4]. The NTSB found that at around 1:19 pm a minor automobile accident occurred on the interstate resulting in a traffic backup. Shortly after this initial accident a driver operating a truck with an empty semitrailer did not react to the traffic backup and collided with the rear of a Land Rover sport utility vehicle (SUV). The SUV was pushed forward into a Hyundai Sonata. The truck then overrode a Kia Spectra and struck a Ford Windstar minivan. The mass of vehicles crashed into a Ford pickup truck towing a trailer and a Chevrolet Tahoe SUV. The accident resulted in 10 deaths among the vehicles' passengers. Five others and the semitrailer driver were also injured. Fig. 17.1 shows a photo from the NTSB report. From the initial impact to the final resting point the truck–tractor traveled approximately 270 ft. pushing and dragging other vehicles along with it. The truck-tractor was traveling at approximately 69 mph, under the 75 mph speed limit [5].

The truck driver had a valid driver's license without restrictions. His driving record showed no violations, convictions, or accidents. Tests after the accident ruled out drugs and alcohol as being involved. Two other tractor–trailer

■ **FIGURE 17.1 Accident photo.** *(Sources: Oklahoma State Police; http://www.ntsb.gov/ investigations/AccidentReports/Reports/HAR1002.pdf.).*

trucks were traveling the same stretch of the interstate at the same time and were able to successfully stop their vehicles.

As part of the investigation, the NTSB learned about the driver's schedule. Specifically they found:

- The driver was regularly scheduled to work on Mondays, Thursdays, Fridays, and Saturdays;
- He generally came on duty between 2:00 and 3:00 am;
- He generally went off duty before 3:00 pm;
- His 17 mile commute to and from work took about 20–30 min;
- He worked a total of 5 days in Jun., with days off for sick leave and scheduled vacation time;
- He worked Jun. 22 and took Jun. 25 as a holiday;
- He returned to work on Jun. 26, the day of the accident.

The NTSB investigated this information to gain an understanding of the driver's potential for sleep debt and how his circadian rhythms may have been influencing him. Ideally, the NTSB would develop a full 72-h history prior to the accident. The truck driver declined to speak to the NTSB investigators but they were able to piece together information from medical consultations and cell phone records. The driver stated that he typically would try to go to bed about 8:00 pm and rise about 12:30–1:00 am. The driver stated when he was not driving his typical schedule he would go to bed around 10:00 pm and rise about 6:00 or 7:00 am. People who spoke to the driver in person or on a cell the day of the accident did not report noticing any sign of illness or fatigue. The driver had been on duty for 10 h continuously when the accident occurred [5].

A requirement for commercial drivers is a periodic medical examination. For the 3 years prior to the accident, the driver consistently denied having symptoms associated with obstructive sleep apnea (OSA). He had checked "no" a medical form for the question concerning "sleep disorders, pauses in breathing while asleep, loud snoring" for the 3 years prior [5]. The driver was hospitalized 3 weeks prior to the accident with complaints of weakness, diarrhea, and abdominal pain. The NTSB report stated, "The hospital discharge summary stated that the driver self-reported "excessive daytime tiredness and loud snoring at night, raising the possibility of OSA (obstructive sleep apnea)" and that he "prefers to not have evaluation of OSA." The NTSB reported that after the accident, doctors noted no evidence to suggest seizures. More than 1 month after the accident the driver underwent a sleep study. His self-reported sleepiness resulted in a score 13 on the Epworth Sleepiness Scale, which indicates abnormal daytime sleepiness [5].

As a part of the investigation the NTSB also reviewed the operations of the trucking business. They operated 7 days a week, 24 h per day. Three times a year, drivers would bid for shifts based on seniority and drive that schedule for 4 months. This work assignment process minimized shift changes and the associated negative impact on drivers' sleep patterns. Driver training was also reviewed. An industry video on sleep, fatigue, and rest was shown to drivers. The company did not have a formal FRMS. The company had a policy that prohibited the uses of cell phones while driving. The driver in this accident violated the policy and used his phone while driving. Previous inspections of the company's trucks and drivers' records were also examined by the NTSB [5].

The NTSB determined the probable cause of this accident was the truck driver's fatigue. The report stated this was caused by acute sleep loss, circadian disruption, and "mild" sleep apnea. As result of this fatigue, he failed to react to traffic conditions. The NTSB identified the structural incompatibility between the large truck and the passenger vehicles contributing factors to the severity of the accident.

The report concluded with recommendations for various stakeholders [5]. Part of the recommendations to the Federal Motor Carrier Safety Administration (FMCSA) included:

> *Create educational materials that provide current information on fatigue and fatigue countermeasures and make the materials available in different formats, including updating and redistributing your truck-driver-focused driver fatigue video; make the video available electronically for quicker dissemination; and implement a plan to regularly update the educational materials and the video with the latest scientific information and to regularly redistribute them.*

To the trucking company, Associated Wholesale Grocers, Inc., they recommended:

> *Create and implement a comprehensive fatigue management program using existing sources of information, and develop a systematic process to update the program as more guidance becomes available.*

The NTSB report also reiterated previously issued recommendations to the FMCSA that had been made relative to fatigue-related accidents and FRMS. As stated previously, there is one more critical step in the accident investigation process. The NTSB, the company, and other stakeholders should ensure these recommendations are followed and the risks associated with tired drivers are reduced. If this step is not performed, the investigation is for naught.

A major goal of any accident or incident investigation should be to recommend improvements to the existing system. An additional improvement that could be made but was not listed by the NTSB is a change to company policies. The driver was able to check a box on a medical form and decline testing for sleep apnea. This disorder was a major cause of the accident that killed 10 people. The risk was identified during the driver's hospitalization but corrective action was not taken. A recommendation to further explore this process breakdown is appropriate. Such a change could not only improve the company's FRMS but also reduce its liability.

17.5 BP TEXAS CITY CSB INVESTIGATION

On the afternoon of Mar. 23, 2005, the BP Texas City Refinery suffered one of the worst industrial disasters in recent times. Explosions and fires killed 15 people and injured 180, caused tens of thousands to shelter-in-place, and resulted in financial losses exceeding $1.5 billion. The incident occurred during a startup process when liquid in a splitter tower was overfilled. As the tower was overfilled, a pressure relief device opened, resulting in a flammable liquid geyser from a blowdown stack that was not equipped with a flare. The flammable liquid was ignited and an explosion and fire followed. The individuals who were killed were in or near office trailers located nearby. Houses were damaged as far away as three-quarters of a mile from the refinery. Due to the scope of the accident, the CSB investigation board not only considered factors at the Texas City facility, but also the role played by BP management in England. The CSB further examined the effectiveness of OSHA as the federal government oversight agency responsible for worker safety. The CSB investigation was conducted in a manner similar to the Columbia Accident Investigation Board (CAIB) examining the loss of the space shuttle. This resulted in an evaluation of both technical and organizational causes for the Texas City incident.

The CAIB explained the need for this type of broad investigation [6]:

> *Many accident investigations make the same mistake in defining causes. They identify the widget that broke or malfunctioned, then locate the person most closely connected with the technical failure: the engineer who miscalculated an analysis, the operator who missed signals or pulled the wrong switches, the supervisor who failed to listen, or the manager who made bad decisions. When causal chains are limited to technical flaws and individual failures,*

the ensuing responses aimed at preventing a similar event in the future are equally limited: they aim to fix the technical problem and replace or retrain the individual responsible. Such corrections lead to a misguided and potentially disastrous belief that the underlying problem has been solved.

Investigators from the CSB arrived at the Texas City facility the day after the accident. They reviewed over 30,000 documents, conducted 370 interviews, tested instruments, and assessed damage to equipment and structures in the refinery and surrounding community. Data from the computerized control system and 5 years' worth information from previous startups were examined. Experts in blast damage assessment, vapor cloud modeling, pressure relief system design, distillation process dynamics, instrument control and reliability, and human factors were involved in the investigation. As part of the investigation a detailed timeline was construction and logic tree causal analysis was performed [3].

The CSB's final report contains findings in both technical and organizational areas. They made recommendations for BP's executive board, the Texas City facility, OSHA, the American Petroleum Institute, United Steelworkers International, and Center for Chemical Process Safety. Using this wide view approach to the investigation CSB identified recommendations in the area of safety culture that go well beyond the specifics of what happened that day in Texas. Their report was a catalyst to the chemical industry's efforts to managing human fatigue risks.

BP TEXAS CITY FINES
OSHA cited BP for a then-record $21 million in 2005 after the Texas City explosion that killed 15 workers. In a 2009 follow-up investigation, OSHA found some improvements but issue 270 failure-to-abate notices, 439 new willful violations, and new record $87 million in fines. By 2010, BP had paid the entire $50.6 million penalty for the uncorrected violations. By 2012, OSHA and BP had resolved 409 of the new citations issued 3 years earlier and paid another $13 million in penalties.

REFERENCES

[1] Inspectors toolkit: human factors in the management of major accidents, health and safety executive. Available from: http://www.hse.gov.uk/humanfactors/topics/toolkit.pdf; 2005.

[2] Guidelines for investigating chemical process incidents. Center for Chemical Process Safety of the American Institute of Chemical Engineers, 2nd ed., New York, New York, 2003.

[3] Chemical Safety Board. Texas City Final Report, Report No. 2005-04-I-TX. Available from: http://www.csb.gov/assets/1/19/CSBFinalReportBP.pdf; 2007.

[4] http://www.ntsb.gov/investigations/AccidentReports/Reports/HAR1002.pdf

[5] National Transportation Safety Board. NTSB/HAR-10/02 PB2010-916202 Accident Report: Truck–Tractor Semitrailer Rear-End Collision into Passenger Vehicles on Interstate 44 near Miami, Oklahoma June 26, 2009. Available from: http://www.ntsb.gov/investigations/AccidentReports/Reports/HAR1002.pdf; 2010.

[6] Columbia Accident Investigation Board (CAIB) Report, vol. 1, Washington, DC: National Aeronautics and Space Administration. Available from: http://spaceflight.nasa.gov/shuttle/archives/sts-107/investigation/CAIB_medres_full.pdf; 2003.

Index

Printed in the United States
By Bookmasters